STAR-CROSSED ORBITS

Other Books by James Oberg

Famous Spaceships of Fact and Fantasy (with Harold Edmonson)

Red Star in Orbit

New Earths

Mission to Mars

UFOs and Outer Space Mysteries

New Race for Space

Pioneering Space (with Alcestis Oberg)

Uncovering Soviet Disasters

History of Orbital Rendezvous (editor)

Space Power Theory

2001: A Mars Odyssey

STAR-CROSSED ORBITS

*Inside the U.S.-Russian
Space Alliance*

James Oberg

McGraw-Hill
New York   Chicago   San Francisco
Lisbon   London   Madrid   Mexico City   Milan
New Delhi   San Juan   Seoul   Singapore
Sydney   Toronto

Library of Congress Cataloging-in-Publication Data

Oberg, James E.
    Star-crossed orbits : Inside the U.S.-Russian space alliance / by James
Oberg.
        p. cm.
    Includes bibliographical references.
    ISBN 0-07-137425-6
    1. Astronautics—United States.   2. Astronautics—Russian (Federation)
I. Title.

TL789.8.U5 O2423 2001
629.4'0973—dc21

2001045262

McGraw-Hill

A Division of *The McGraw-Hill Companies*

1 2 3 4 5 6 7 8 9 0   AGM/AGM   0 7 6 5 4 3 2 1

0-07-137425-6

Printed and bound by Quebecor / Martinsburg.

This book is printed on recycled, acid-free paper containing a minimum
of 50% recycled de-inked fiber.

For my parents, John and Jean Oberg,
who taught me to love truth

Contents

Acknowledgments

Aerospace technology is one of the few arenas of human activity where you can't bluff or bully your way through, or camouflage shortcomings with a blizzard of excuses and rationalizations. Although its practitioners strive for among the most challenging and exciting of endeavors, to overcome (or at least evade) gravity, they realize these dreams must be founded on severe reality. Anything that deflects or distracts from reality is 'bad', and can instantaneously exact a horrifying price from the careless. As the bumper sticker says, "Man forgives, God forgives, nature never."

So the first acknowledgment of this book must be to the men and women, of all nations and of all times, who have dedicated themselves to this profession and to the ferocious obsession with truth that its proper pursuit entails. Their accomplishments will define the way our era is remembered in times to come, and what they have accomplished will be remembered when their names, and the names of politicians and movie stars and sports champions, and the names of countries themselves, have all been forgotten. Without the wisdom, wealth, and power that space activities are making available, I doubt that our civilization, our species, even our entire biosphere, can long endure.

Second only to this is the acknowledgment of the help of hundreds of these aerospace professionals in collecting, sharing, and assessing the information in this book. Since they recognize that preferring reality over make-believe is not a mental trait widely shared, and is often punished, they know that their cooperation with me must be anonymous, for their own protection, except in special circumstances when they can allow their names to be used. This places a tremendous burden on me, to demonstrate to my readers the credibility of statements from people who are unavailable to verify these statements. All I can do is assure my readers that I know and trust the people whose information I've used.

A book like this is just "the tip of the iceberg" of the pursuit and promulgation of an accurate reconstruction of events and usable deductions from them as a guide to the future. This is an activity that has influences on my personal life both during working hours and beyond. So I owe an acknowledgment to my family—my wife Cooky and my sons Greg and John—for their toleration of the distractions from my life with them.

Making this body of material into a potentially coherent book requires knowing where to start, where to stop, what to keep, and what to leave behind. The insights and sound judgment of my editor Amy Murphy made this possible, and it was a delight to have her creative and constructive advice on this. Critical editing from Natalie Freeborg was also indispensable. The guidance from my editing supervisor, Ruth Mannino, in getting this material from manuscript to bound book was invaluable. Other associates of mine, who cannot be named for a variety of reasons, helped critique particular portions of the text. Considering the quality of these reviews, all remaining garbles and flubs, and all careless oversights of any others who helped me on this book, are entirely my responsibility.

James Oberg

STAR-CROSSED ORBITS

Introduction

"Convictions are more dangerous enemies of truth than lies."

Nietzsche

"The greatest obstacle to discovering the shape of the earth, the continents, and the oceans was not ignorance but the illusion of knowledge."

Daniel Boorstin

The proverbs of the old Appalachian Mountains contain wisdom that is too often missing in modern life. One in particular is useful for people struggling to understand the issue of Russian and American space activities. Popularized by Will Rogers, it goes like this: "It ain't what you don't know what'll make you look like a fool, it's what you *do* know what ain't so."

In 1988, as a guest on CNN's *Crossfire*, I was actually impudent enough to make use of that proverb on national television. A two-man Soviet space capsule appeared to be stranded in orbit, and I was a member of a "panel of experts" discussing the crisis. I listened as my fellow guests—the Defense Intelligence Agency's former director, Lieutenant General Daniel O. Graham, and New Jersey Congressman Robert Torricelli—expounded on the impending doom in space.

"They're dead men already," Graham asserted, describing how their air would soon run out. Even if they did return to Earth, it would be tough. "The *Soyuz* smashes to Earth with terrific force," Torricelli expounded. Both speakers, I knew, were men of intelligence, experience, and integrity, so I was astounded at how preposterous their comments seemed.

When my turn to speak arrived, I quoted the proverb and then described the variety of tricks that the crew still had at their disposal to make a safe (and soft) landing. I expressed my willingness to "bet the farm" that they'd be safe on Earth swapping yarns with rescuers within hours (as indeed they were). My fellow panelists shook their heads in disbelief.

Later on, when somebody on Torricelli's staff explained to him what (or who) the proverb I had quoted had actually been referring to, he reportedly fired off an angry note to CNN about my "disrespect." Still later, I learned what none of us had known at the time: that the cosmonauts had actually been within seconds of an irrevocable death sentence. Unbeknownst to the entire outside world, they had dodged it, but Graham and Torricelli's instinctive pessimism was closer to the reality of the situation than I had wanted to believe.

At the time, there was a lot we didn't know about Russian space activities. There was a lot we all thought we knew that wasn't so. There was a lot that some people wanted us to believe, whether it was true or not. Worse, there was a lot that specific people didn't want us to know. And most frighteningly, as the fates of the American and Russian space programs became increasingly intertwined throughout the 1990s, there turned out to be a lot that many people wanted *not* to know, lest it get in the way of the "politically correct" policies that NASA's managers had established in response to directives from the White House.

Space is empty, but it is not a blank slate. Travelers there must carry their own air, food, and water, but they also carry the heavy burden of their history. They carry with them what they know, or think they know, about each other. So for the last half century, as Russia and America pioneered the space frontier beyond the physical boundaries of their home planet, they interacted with each other in a context that was rooted firmly back on Earth. The Russian and American space programs have been linked together since the 1950s, and over time, the nature of the relationship has changed dramatically. It has been by turns inspirational, sloppy, wasteful, and politically and even scientifically useful.

Historians have come to believe that this relationship was a natural outgrowth of the past. Some believe that it was the only feasible way to create and sustain Earth's energetic exploration of space into the future. But its critical value hasn't always been obvious. And its true dynamics remain obscure today. Just look at all of the surprises and disappointments in the current "space partnership" between the United States and Russia. With almost half a century of space diplomacy behind us, how could we still be so myopic and clumsy in our dealings with the Russians? And how could they so regularly misinterpret our intentions in space?

Back in the mid-1950s, just before the dawn of the Space Age, philosophers and science fiction writers were busy trying to forecast the social forces that would motivate the tremendous commitments of resources that would be needed to make it happen. Clearly, space exploration was going to be expensive. The question was, where would the money come from?

I remember the speculations. Already a "space nut" as a boy, I read everything I could lay my hands on. I watched the pre-space-flight science fiction shows on television; I listened to space programs on the radio. The Space Age was about to be born, but who would be the midwife? Nobody even knew who the father would be.

For generations, fiction had invented weird and wonderful motivations for space travel. Jules Verne's "Baltimore Gun Club" sent men around the Moon in order to demonstrate their advanced military skills. H. G. Wells's Martians crossed interplanetary space to flee a dying planet, and sequels by other writers portrayed human fleets (with electric engines designed by Thomas Edison) headed for Mars on a mission of revenge. Contemporary with Wells, the Russian space-flight prophet Konstantin Tsiolkovskiy imagined humans inspired by a "life force" expanding into a new ecological niche.

Some writers in the 1950s thought that commercial ventures would seek profit in space. Others suggested military parallels between atomic energy (brought on by World War II) and space travel (perhaps created during a future U.S.-Soviet war). Almost every theme in the history of terrestrial exploration, from the Vikings to the conquistadors to the Hanseatic League to the Mormons, was reset in the black emptiness of space. Science and general human curiosity received passing mention as well. A young writer named Ben Bova tried to sell a short story about space exploration sparked by U.S.-Soviet diplomatic competition, but his editor rejected the notion as too unbelievable.

Then, on October 4, 1957, to the great shock of the entire world, *Sputnik* was launched. The Space Race was born, and it took nearly everyone—experts as well as ordinary citizens—by surprise. Surprising, too, were the compelling reasons that soon arose in its support.

Hindsight has allowed historians to recognize why Nikita Khrushchev authorized his rocket builder, Sergey Korolyov, to divert one of his new intercontinental missiles from the Soviets' weapons testing program. The missile would be used to carry an artificial satellite into orbit around Earth. Scientific curiosity, or even Soviet grandstanding, was far from his mind. He needed to prove to America that he could kill millions of civilians in the event of a war. And, faced with a crushing military budget, he needed to prove to his own generals that there were millions of unneeded soldiers in the Red Army.

Once the worldwide (and internal) admiration for Soviet space successes became evident, new goals were developed. A solid foundation for space development was not on the agenda, however. The idea was to go for the "firsts" that would garner headlines around the world. And it was those very headlines that endured, even after the crippling effects of shortsighted goals became evident. The consequences of the hollow triumphs that appeared in those headlines are still shaping our attitudes and beliefs today.

In the West, meanwhile, there was genuine fear that an efficient, centralized Soviet space industry would capitalize on its lead and irreversibly dominate all scientific, commercial, and military aspects of space flight. The shocking reality of the Soviet Union, the feudalistic and byzantine hodgepodge of industrial empires slipping back and forth between unsteady alliance and outright hostility, was to lie hidden for decades. From the very beginning of the Space Age, the Soviets sought to develop space weapons, both automated and manned. They justified their aims by fears of American military plans. Indeed, there was no shortage of American experts calling for the construction of just such weapons, and no lack of industrial organizations eager to build and sell them. In public, however, Soviet spokespeople—along with a broad swath of Western experts—denounced any American moves toward military space activities. At the same time, they minimized or totally denied the existence of Soviet plans in that direction. Some Pentagon statements during the Reagan years looked pretty paranoid, but in hindsight, these fears turned out to be justified and even incomplete. Such fears would begin to fade only when the Soviet Union itself had ceased to exist.

Deliberate delusion continued. With the Moon race over, the Soviets denied the "American victory," claiming that the Soviet Union had never actually been serious about competing. Nor was self-delusion absent. A few politically motivated joint projects in the 1970s convinced many people that "cooperation in space" could actually change the attitudes of political leaders back on Earth, through pure force of example.

The relationship has since evolved into new patterns of mutual misunderstanding. From the joint near-bankruptcy of the early 1990s to the enthusiastic "marriage in the heavens" of mid-decade, we've become disillusioned even as hard-won joint achievements have appeared in orbit. As Ben Franklin advised about marriage, people should go into it with their eyes open, but keep them half closed afterward. But with the Russians in space, we seem to have done it backward.

Over the last half-century, the dynamics of the Russian-American space competition have powered the space programs in both of these countries and in dozens of their allies. Any consideration of how this has happened, and what forms this relationship will take in the twenty-first century, needs to understand the nature of the "engine."

One widespread, overwhelming, and fundamental misconception about this relationship still stands in the way of understanding it and successfully predicting future trends. Space activities in one country, we are told, influence the planning of space activities in the other country, and vice versa. This is what engineers call a "feedback loop."

In practical terms, however, this is not true now and never was. Mistakenly believing it to be true has led the space programs in both countries to unfounded fears, enthusiastic wild goose chases, time-consuming detours, and heartbreaking setbacks, all at great expense.

What, then, *does* drive this feedback loop? What foreign forces influence a nation's policies? We only have to look back on the dynamics of the past few decades to see the real driving principle.

It turns out that each country's program is not directly influenced by what is truly happening in the other country's program. Rather, it is influenced by what each country's leadership *thinks* is happening in the other country. This is so obvious that it is usually never stated explicitly. As a result, when one country's perception of the situation in the other country is flawed, the response will necessarily be misdirected and mistaken.

Early in the Space Race, the gap between what was real and what was perceived was usually the result of the uncertain nature of the technology and the tendency of each side to attribute its own motivations (or fears) to the other side. Sometimes the worst case had to be prepared for in the absence of reassuring knowledge. Sometimes judgments were colored by what decision makers felt comfortable "knowing."

Often these judgments were derived from misconceptions about the nature of space or were accepted because they satisfied preexisting attitudes. There was the "crude but powerful Russian rockets" myth, for example, which acknowledged that the Soviets' vehicles were big but belittled them for being "unsophisticated." There was the persistent belief that the USSR had concealed the deaths of a dozen "secret cosmonauts," matched by the USSR's undying anxiety about America "militarizing space" (and to mirror that, the American fear of "Soviet bombs hanging over our heads").

In other situations, advocates of a minority position in one country would attempt to strengthen their domestic position by interpreting the other country's program in terms consistent with the advice that they themselves had given. If you wanted your country to build a space station, or a space shuttle, or a space laser, it helped if you could make it look as if the other country was doing this already. And if you couldn't get the key officials to agree, you could bypass them in the national news media (at least in the United States).

That was the stage on which I played the public role of a "space sleuth." Since the early 1970s, I had been publishing articles and giving news interviews on what I thought was the truth behind all the claims and misunderstandings. Sometimes I would work on a particular puzzle for years; sometimes I had to give off-the-cuff interpretations and predictions.

Along with a loose confederation of like-minded private space sleuths around the world, I believe I made many contributions to public understanding. Together, we published some pioneering insights that stood up pretty well in hindsight. I had my share of near misses, and some outright strike-outs, too. But over the years, I developed some helpful techniques, as well as some valuable resources, in my quest "to find out, and to tell about" (my unofficial slogan) space-flight mysteries.

Then *glasnost* appeared in the USSR, and a torrent of restricted information poured forth. After the collapse of communism in 1991, even greater access to people, old hardware, space factories, and archives became possible. For a while, it looked as if the best verdict I could expect from history was to have been that guy who used to be able to guess really well about the Soviet space program, way back in those dim times before we didn't have to guess any more.

That obituary for a lifelong avocation was premature. The Russians soon had an entirely new array of motivations for keeping "space secrets." As Russia and the United States moved closer in formal cooperative projects, my techniques and resources became more important than ever. The fate of America's space program had become dependent on reliable knowledge of those Russian space activities that were directly involved with American projects.

This is the story to be told in this book. Yet considering in all humility how incomplete and observer-dependent many of these "realities" really are, the concept of an omniscient narrator is laughably inappropriate here. Instead, imagine the narrator as a tour guide. This will be a personal journey of discovery and understanding, helping us to find out just what it is possible to "know" with confidence. It will also involve defining the boundaries of the known, and speculating about what remains to be discovered and what has already been lost forever.

Let's start with a specific example. It involves a discovery about the past that has persistent relevance to our present and our future. It's funny, sad, and frustrating all at once. And it's so typical of what we carelessly "know" about each other's space activities that was never really so.

Going through other people's attics is one of life's vicarious pleasures, considering the treasures you might stumble across. For me, going through Russian space attics is doubly so, because of the genuine treasures that you can stumble across and the intriguing new patterns that the new data points fall into. In the basement of an auction office in New York City in 1997, I gained a new appreciation for the scope of what had been an old, vague pattern.

As the Soviet Union racked up one "space first" after another in the 1960s, it also performed the bureaucratic duty of registering many of these firsts. The world body responsible for all flight records was the International Aviation Federation (FAI, from the organization's name in French) in Paris. Registrations with the FAI took the form of bound, large-format descriptions of the events for which the claims were being made, with appropriate official signatures. One of the most frequent signers for the Soviet claims was Ivan Borisenko, titled "sports commissar."

Just why he had been chosen I could never figure out. Maybe he owned the stopwatch. In any case, a few decades after the Soviet-era glory came the post-Soviet cold and hunger. Struggling on with an inadequate pension, Borisenko produced his own personal archive of two dozen space record claim folders and offered them for sale in the West.

It was this set of handsomely bound documents that I was inspecting and authenticating for my host and paying client, Kaller's America Gallery. We would catalog each one, and I would read it over in Russian to note the accuracy of its claims. One thing that I noted about the claims was the almost universal insistence that the launch site of these "space firsts," Baykonur, was located at precisely 47°22'00" north, 65°29'00" east. Ever since the first American U-2 spy plane flew over Russia in 1956, however, the launch pad has been known to be at 45°55'00" north and 63°20'00" east. Foreign observers had always suspected that the error was deliberate, presumably to get the next U-2 spy planes to stray off course. Finally, in an incredibly rich collection of Russian space memoirs published in the same year as the auction, two former Soviet officials independently described how the falsehood originated. It was just as we suspected, but it's the real inside story.

Vladimir Yastrebov, an expert in spacecraft tracking, wrote about his exact role in the deception: "I was personally involved in naming the Tyura-Tam launch site 'Baikonur' so as to disguise its true location. A few days after Gagarin's flight, my management sent me to one of the central administrations of the Ministry of Defense to meet with Col. Kerim A. Kerimov [the officer in charge of the cosmonaut program]. Together with a senior officer from his section called Alexei Maximov, I was asked to draw up the records of Gagarin's flight in terms of range and altitude for registration with the International Aviation Federation in Paris. Preparing the document was easy enough, but we encountered a major hurdle when deciding how to identify the site from which the *Vostok* launch vehicle had lifted off. Since we were not allowed for security reasons to name Tyura-Tam, we studied the map and chose a ballistically plausible down-range alternative in the form of a small Kazakh settlement called Baikonur. And that is what the cosmodrome has been called ever since."

Reading further in the same book, *Roads to Space*, I found that Aleksandr (not Aleksey) Maksimov, an official of the Ministry of Defense responsible for space activities, had also contributed a memoir. He told much the same story, but slightly garbled with regard to the dates and organizations: "So where did the name Baikonur come from?" he wrote. "In accordance with an international treaty, we had to register our Aug. 21 [1957] ICBM launch with the United Nations, indicating the date, time, and place of launch.

"Since there were no spy satellites in orbit yet, nobody knew where the test range was situated, and we were not keen to divulge that information for security reasons. We therefore decided to indicate a site whose existence the Americans could verify. With their radars they were able to track the flight of our rocket and, by working backward, calculate the approximate location of the launch site. So we decided to give the Telegraph Agencies of the Soviet Union [TASS, the main news agency] and the United Nations the name of a place situated some 250 kilometers from Tyura-Tam. That place happened to be called Baikonur—and ostensibly that is where we have been launching from ever since."

Yastrebov's account is more accurate, since the Baykonur story was associated with the first manned flight aboard *Vostok* and with the 1961 FAI registration, not with the earlier missile test. But Maksimov's account is essentially corroborative regarding the motivation and the action itself.

So the official claims contained intentional falsehoods. I'd always presumed that the FAI had prohibitions and penalties for submitting knowingly false claims, and there can be no question but that these data were submitted in full knowledge that they were false. Nobody expected the Soviet Union to tell the truth, so we all became accustomed to swallowing lies. In recent years, however, Russia has wanted to become a normal country, to behave by internationally accepted norms, and to earn the trust of the world. Could standards be applied retroactively?

Sure enough, I found the FAI "Sporting Code" on the Internet. It has an entire section on "Complaints," and section 5.2 is entitled "Penalties and Disqualifications." Subsection 5.2.2.3 defines "Unsporting Behavior" this way: "Cheating or unsporting behavior, including deliberate attempts to deceive or mislead officials, falsification of documents, or repeated serious infringements of rules should, as a guide, result in disqualification from the sporting event."

There was no need to withdraw the flight records, since the Soviets really did perform the feats described. But I was hopeful that the false information could at least be expunged from the archives of the world body. I figured that the best way to do that was to have some official ask the Russians to file a letter of amendment to the original claims.

It wasn't as easy as all that, I discovered. I located the U.S. association affiliated with the FAI, the National Aeronautic Association in Arlington, Virginia, and I proposed to them that the Russians be asked to correct the false information on their original records claims.

On November 21, 1997, association official Art Greenfield (the secretary of the Contests and Records Board) wrote back to me to politely explain why that wasn't going to happen. "I understand that you believe the Russians falsified the coordinates of the launch site of those flights in the record dossiers," he began, adding that since the association didn't have the dossiers on file at its office, he had no way of confirming this.

"Perhaps the Russians did attempt to mislead us about the takeoff location for reasons of national security," he conceded. However, since the actual flights are not in doubt, "we see no compelling reason to confront our Russian counterparts with allegations of wrongdoing dating back to the Cold War era." He concluded by saying that these days, both Russians and Americans "are actively involved" with work that "promotes public understanding and awareness of the importance of space flight," and furthermore, that "we hope that this cooperative effort will continue for as long as we explore space."

Max Bishop, the FAI secretary general in Paris, concurred. "No space records depend on the precise location of the launch site," he pointed out, quite correctly. "Therefore modifying the coordinates of Baikonur will in no way affect any FAI-approved performance. We do not intend to take any action."

Perhaps that's the proper perspective. After all, it is reasonable to question the importance of a 1961 fraud in 2001. That is, is there anybody out there who doesn't already know that the official Soviet location for the cosmodrome is false? Why bother with an official correction?

A compelling reason is that the original deception persists through sheer informational inertia. Even a cursory survey of existing cartographic products shows this. For example, recent world globes from Replogle (such as the World Horizon "Livingston Illuminated" globe) and a World News Map published by *U.S. News and World Report* show the town of Baykonur in its correct location. But I would argue that nobody looks up Baykonur out of interest in obscure coal-mining towns (in population and genuine importance, it's much too minor a spot to earn its own place on these maps). People look up Baykonur because they want to find out where the famous cosmodrome of the same name is located. If so, they are misled, since it is the erroneous assumption that the cosmodrome is located at the "false Baykonur."

So I play this game whenever I visit bookstores, and you can play too. Check out the latest world atlases to see if they have the cosmodrome at the correct location, on the Syr Darya River just east of the Aral Sea, or if they put "Baykonur" where the original and utterly unimportant town still is. Hammond's *New Century World Atlas* (1997) has the false location, as does *Webster's Concise World Atlas* (1998). So does Rand McNally's *Classic World Atlas* (1996). The French mapmaker Gabelli issued a map of Asia in 1994, and it showed the false Baikonur.

Even more explicitly, the 1994 *Oxford Encyclopedic World Atlas* has a special updated section on the new post-Soviet geography. Its feature on Kazakhstan specifies the Baykonur Cosmodrome as one of the most important features of that new country. But the Baykonur shown on the actual map is the deceptive one. And in the *Oxford Dictionary of the World*, the definition of "Baikonur" on page 63 is, "a coal-mining town in Kazakhstan, n.e. of the Aral Sea. Nearby is the Baikonur Cosmodrome." Neither the Oxford atlas nor the other misleading products show anything at all near the Syr Darya River, where the cosmodrome and its support city of Leninsk are actually situated.

Some do get it right, such as *National Geographic*. Some list the old "Baikonur" but also have correctly located entries such as the "Space

Launching Centre" or "Leninsk" (the city where the space workers live). But they obviously didn't rely on official FAI documents for their information.

Without making too big a deal out of a minor historical falsification, I've always figured that continuing to tolerate such deception is an insult to modern Russia. Isn't it just a condescending way of saying, "We know Russians are liars, so why bother to expect them to tell the truth?" If I were Russian, I would deeply resent such bigotry.

This isn't just ancient space history. The same attitude has persisted all the way into current times. Throughout this book, we shall see many cases in which American officials talk themselves into tolerating Russian deception, since, after all, "they're only Russians" and we need to get used to it. I will argue that we have reaped a frightful harvest from our carelessness toward truth.

There were other old distortions that had alarming implications, whether they were deliberate or the result of ignorance. I remember how Soviet cosmonauts joined Moscow's propaganda campaign against the U.S. shuttle program in the early 1980s, viciously accusing it of all sorts of space weapons activities. The accusations were false—I was working at Mission Control during those early missions and knew exactly what was and wasn't being done aboard *Columbia*—and I confess to harboring unkind thoughts toward the cosmonauts and what I then assumed were their deliberate lies.

I collected dozens of examples. In August 1983, Georgiy Grechko told a television audience, "We know that sights for laser weapons have already been tested on the first shuttle craft." Vladimir Shatalov, head of cosmonaut training, stated, "We Soviet people, in particular cosmonauts, are pained to hear that some people in the United States are trying to use space for military purposes." Aleksey Leonov proclaimed that "the Soviet side repeatedly underscored the fact that space must never be allowed to be used for deploying weapons." Valeriy Bykovskiy denounced the "insane plans" of U.S. militarists "who want to rule the Universe." And according to another cosmonaut, Vladimir Dzhanibekov, the U.S. shuttle "responds to a great extent to the interests of war, not peace."

At the time, I was concerned that such a campaign might be a prelude to Soviet military action against a shuttle mission. Most alarming was a statement by cosmonaut Georgiy Beregovoy concerning the right to orbit in space over other countries. He said that it was not unconditional; it was dependent on peaceful intent. I worried about the implications of a Soviet move to deny that right to American space shuttles.

But in hindsight, I've begun to suspect that these shrill accusations were actually aimed at the internal Soviet debate over whether to preserve the USSR's own expensive "shuttleski" program, its effort to duplicate NASA's shuttle program. This program had originally been sold to Soviet leaders as the answer to a perceived military threat from the American shuttle. Hence, it might lose support if the Soviet domestic perception of the threat were allowed to fade. This could have accounted for the accusations. The Soviet statements could have reflected the Soviets' entrapment in their own sincere misperceptions of American intentions.

The consequences of these mutual misunderstandings went far beyond the space arena. Space scientist Paul Spudis recently argued that they may have been critical to the end of the Cold War.

Writing about the *Apollo-11* lunar landing on its thirtieth anniversary in 1999, Spudis put Kennedy's choice of a "Space Race" finish line into perspective: "The goal of the Moon was a technological challenge, a gauntlet thrown down before our global competitor, the Soviet Union, challenging them to a technological fight to the finish. Although it is commonly acknowledged that we won this challenge, the profound effects of that victory are less often considered."

The Soviets attempted to beat the United States to the Moon. They built rockets, landing craft, and space suits for the purpose. They trained a generation of cosmonauts, flight controllers, and scientists for the purpose. But their efforts failed, even as *Apollo* triumphed.

"What lessons did the Soviet Union draw from this disaster?" Spudis asked. "Apparently, the Soviets became convinced that, in programs of vast technical scope, particularly those requiring the practical application of high technology to very complex problems, America could accomplish anything. The Soviets viewed the Americans as having achieved, through a combination of great wealth, technical skill, and resolute determination, an extremely difficult technological goal." And they never forgot it.

Twenty years later, another U.S. president laid down another technological goal. This time, Ronald Reagan challenged the American aerospace industry to "render nuclear missiles impotent and obsolete." He called the project the Strategic Defense Initiative; critics derisively dubbed it "Star Wars." But the criticism came from the same sources—sometimes the same individuals—that had told Kennedy that the *Apollo* lunar program was foolish and futile.

Soviet officials campaigned against Star Wars for years. "Why did the Soviet Union fight so long and adamantly against it?" Spudis asked. "Clearly the Soviet Union was convinced the SDI would work and that we would achieve exactly what we set out to do."

"Here's *Apollo's* legacy," he argued. "Any technological challenge America undertakes, it can accomplish. The reason this legacy had currency was the success of *Apollo*. We had attempted and successfully achieved a technical goal—one so difficult and demanding, that it made virtually any similar goal seem equally achievable.

"Moreover, this was a goal that the Soviets themselves had attempted and failed," he continued. "They reasoned that getting into a decade-long competition with America on SDI would similarly end in an American victory and would be a race that would bankrupt and destroy their system, as indeed, it did."

In conclusion, Spudis maintained that it was the Soviet perception of *Apollo*, rather than the reality of *Apollo*, that had had profound results: "The success of the *Apollo* program gave America something it did not realize was so important—technical credibility. When President Reagan announced SDI 20 years later, the Soviets were against it, not because it was destabilizing and provocative, but because they thought it would succeed, rendering their vast military machine, assembled at great cost to their people and economy, obsolete in an instant. Among other factors, this hastened the end of the Cold War in our favor."

The reality of a reliable missile defense could easily be called into question, but the actual truth wasn't important. Far more critical was what decision makers in Washington and Moscow believed to be true, or even plausible.

The end of the Cold War and the collapse of Soviet totalitarianism were supposed to make Russia a "normal country," one that didn't need propagandistic deceptions. But these changes did not bring an end to space-related misconceptions. In *Space Policy* magazine's January 2000 issue, veteran international space projects manager Jeffrey Manber published a report called "Russian-American Space Miscommunication: A Study in Missed Opportunities."

"Mired in political and programmatic confusion, American-Russian space cooperation is at a crossroads," Manber wrote. The process is "still critical to the future of cost-efficient space exploration yet grounded by political barriers and misunderstanding.

"Much of the misunderstanding can be traced to the very beginnings of Russian-American space cooperation," he continued. "There were actually two beginnings: the first a commercial effort and the second, the more publicized overture of NASA to work with the Russian space industry. Both have failed on some very basic levels."

In 1988, a small Boston company signed a contract to fly a modest research payload inside the *Mir* space station. Manber described how top Russian space officials misinterpreted this event, probably because they were anxious about diminishing Soviet government funds for expensive space projects.

"To my surprise," Manber wrote, "they believed the export license from the Reagan Administration to be a well-orchestrated signal from the U.S. government. . . . [It] signaled that a commercially structured Russian program, one supported by market conditions, would be met with political and commercial support from the Americans." Manber remembered trying to convince Soviet officials that the license was a fluke, snuck past opposing forces at the State Department and NASA through bureaucratic back channels.

"Unconvinced, the Soviets left the conference still believing that the US would welcome a commercial Russian space program, even if it competed against the US government space station, or the monopoly status of US launch vehicles and, in general, the non-commercial slant of NASA's space programs," he wrote. "Thus the working relationship between the Russians and Americans was born in confusion and misunderstanding. Little has changed since then."

Manber's pessimism may be exaggerated, in large part because of the frustrations he faced in his position on the frontier of many of the cooperative efforts. As the old proverb goes, you can always recognize the genuine pioneer. He's the guy lying face down on the trail in front of you with an arrow in his back. Manber, who has devoted the last decade of his life to painfully advancing U.S.-Russian commercial space cooperation, can empathize a lot with that guy.

Where do we go from here? Once the problem has been recognized, it's remediable. Confusion and misunderstanding must be reduced. Years of hard experience have accomplished this goal among many of the American participants in the current cooperative phase of the U.S.-Russian space relationship. For as long as this phase lasts, and especially when the current phase eventually ends and the next phase begins, we must strive for a relationship based on reality. Common interests and complementary skills must benefit everyone involved, and the effects of mutual misunderstandings and baseless fears must be minimized.

As expensive disasters show again and again, outer space is unforgiving of error and of self-delusion. But as we shall see, this is a lesson that many at NASA willfully ignored.

1

Zenith

"Yet ah! why should they know their fate?
Since sorrow never comes too late,
And happiness too swiftly flies.
Thought would destroy their paradise.
No more; where ignorance is bliss,
'Tis folly to be wise."

Thomas Gray, "Ode on a
Distant Prospect of Eton College"

"There is only one thing more painful than learning
from experience and that is not learning from
experience."

Archibald MacLeish

Konstantin Feoktistov was unhappy. A brilliant yet undiplomatic Russian space official, he was famous for never remaining unhappy alone. He longed to brag about the spectacular accomplishments of his cosmonauts and space engineers in overcoming one of the worst crises in the history of human space flight.

Soviet "space secrecy" had forbidden any public mention of the crisis. The frustrated Feoktistov fumed in private, complaining bitterly to his closest associates. His entreaties to the news censors were rejected.

The reputation of the Soviet Union trumped the reputation of Feoktistov's team, and Soviet space endeavors could in no way be allowed to appear fallible.

It was the summer of 1985, and the Soviet space program was on a roll, with impressive achievements behind it and spectacular new advances only a year or two away. The lost "space race" of the 1960s was a fading memory. Feoktistov and his colleagues must have been thinking that the Americans had better get used to eating space dust.

Yet the idea that this new plateau of space mastery would lead forward to even greater heights was actually an illusion. A handful of triumphs over a few more years would be all that the Soviets would achieve. This was the zenith of their lifelong efforts, the apogee of their space orbit.

It certainly didn't look like it at the time, to either side. While the Russians were jubilant, many Americans were anxious. They were slowly beginning to suspect that a new space race was underway. Fears were mounting that just as in the 1950s, the Soviets were about to grab a commanding lead. True, NASA was operating a sophisticated fleet of science and applications satellites, with probes penetrating far into the Solar System. True, NASA's space shuttle and its *Spacelab* research module had become operational in 1982–1983. And true, President Reagan had just ordered NASA to build a human occupied space station and had simultaneously opened a new front in the Space Race by challenging space engineers to build a workable antimissile system, officially called the "Strategic Defense Initiative" but popularly known as Star Wars. But it looked as though the Soviets would soon match and even surpass American efforts.

At congressional hearings in March 1983, Robert Cooper, head of the Department of Defense research group DARPA (Defense Advanced Research Projects Agency), sounded a warning. "All of that space activity leads one to believe the Soviets have some grand scheme," he testified. "[They] are building components which apparently could come together into several different kinds of new space-launch vehicles. They are building at the same time brand new space-launch complexes that are as large and complex as the ones that we are building at [the Kennedy Space Center in Florida] and [Vandenberg AFB in California] for the shuttle program."

A year later, a leading American aerospace news weekly, *Aviation Week and Space Technology*, expanded on this prediction. "Several large new systems should be operational within five years in a space program build-up as large, if not larger, than the U.S. Apollo effort," the magazine informed its readers. "The Soviets have as many as nine aerospace and heavy machinery design bureaus working on this space buildup."

Indeed, in 1985, a great deal lay ahead for the Soviet space program. The multimodular space station *Mir* would be launched within a year, and a series of add-on laboratory modules were scheduled to follow. The *Soyuz* space taxi, designed to carry teams of cosmonauts up to space stations and back, was being totally redesigned for improved efficiency and reliability. The super-booster *Energiya* was being primed for flight, along with a space shuttle called *Buran*. Supersophisticated science probes to Mars, Venus, and Halley's Comet were already exploring new interplanetary frontiers, and more such probes were on the way. Orbital tests of space weapons—from cannons to lasers to "space mines"—had already occurred, and the development of follow-on systems with greater range and lethality was nearing completion.

Ahead of NASA lay tragedy and frustration. The *Challenger* shuttle would blow up in January 1986, killing seven astronauts. Plans for NASA's *Freedom* space station would be delayed year after year. NASA's most ambitious interplanetary probe in a decade would vanish forever just as it reached its target planet, Mars. Even the "flagship satellites," such as the $2 billion Hubble space telescope, would prove to have crippling flaws.

But as Meryl Streep (playing Karen Blixen, known by the pen name Isak Dinesen) narrated in the film *Out of Africa*, "God made the world round so we could not see too far ahead of us on the road." From space, the world's roundness is plainly visible, yet in space, heartbreaking surprises and bizarre twists awaited the world's space programs.

Feoktistov, who had played a key role in developing Russia's first human piloted spacecraft in the 1960s and had even once traveled into space himself, knew the reality that was sparking American anxieties. By the end of June 1985, after his team's stunning but still secret space triumph, he was sure that it wasn't just hardware that the Americans needed to be worried about.

Described as "a difficult, unsociable, and uncompromising man," Feoktistov had overcome hardships of his own. At the age of 16, he had been a scout for Russian partisan units in World War II. Captured by the Germans, he was shot in the head and left for dead. He recovered from that wound, but later lost several fingers on his left hand. In 1964, aboard *Voskhod-1*, he became the first human to fly into space wearing glasses. Sixteen years later, at the age of 54, he trained to fly into space once again to inspect the *Salyut-6* space station, which his team had built. Shortly before launch, however, he was grounded for health reasons. Over the years, Feoktistov had led the USSR's most talented spacecraft engineering teams in success after success.

Now his space team had just done the impossible. They had sent a crew to an apparently dead space station and resurrected it. A power failure had left the three-year-old *Salyut-7* frozen solid in orbit, drifting aimlessly, with any future missions put on indefinite hold. But an emergency mission to the 20-ton trailer-house-sized orbital outpost had been thrown together in three months. Two veteran cosmonauts had blasted off inside a *Soyuz T-13* spacecraft, docked with and boarded the derelict ship, then diagnosed and fixed the problem and repowered the orbital outpost.

For the Soviets, this repair operation was a breathtaking breakthrough in bold, on-the-fly "contingency" (NASA's term for an unpleasant space surprise) operations. This hadn't been typical Soviet space strategy. During the more than 20 years since their first space missions in 1961, Soviet cosmonauts had followed strictly scheduled space mission scripts, relied on automated equipment, and aborted flights when things weren't going right. And they aborted a great many flights, always claiming in public that all had gone perfectly.

By contrast, NASA built its spacecraft with contingency operations in mind. American astronauts therefore had the flexibility to quickly modify plans and procedures when unexpected situations arose, as they almost always did. This allowed American space missions to overcome problems and complete space dockings, Moon landings, and space station expeditions. Even under the worst conditions, astronauts were better equipped to return from space alive in the event of an emergency. NASA and independent space experts all considered this feature to be the most significant edge that the United States had over the Soviet Union in space activities.

Feoktistov knew that this American edge had been neutralized by the success of his cosmonauts on *Salyut-7*. But he wanted the rest of the world to know it, too. In his campaign to give public credit where it was due, Feoktistov had allies, some in Moscow and some, surprisingly, overseas. Thanks in part to articles published in the West, Feoktistov was given the necessary tools to pry his report loose from Soviet censors.

This space resurrection began, as all resurrections must, with an obituary. On March 1, TASS, the official news agency in Moscow, released a terse and unexpected message about the Soviet space station: "As the planned program of work aboard the *Salyut-7* orbital station has been fulfilled in its entirety, the station is now deactivated and continues its flight in an automatic regime."

This announcement caught Western space experts completely by surprise, since they had been expecting more cosmonaut visits to the station as a matter of course. Launched in 1982, *Salyut-7* had hosted two cosmonauts on a record-breaking seven-month-long expedition. In 1983, equipment problems had frustrated what was to be a second long-duration visit, and the crew had been forced to return to Earth without ever docking to the station. Another mission, designed to practice expanding the station with attachable laboratory modules, had run into a crisis when a tank of explosive rocket fuel leaked overboard through a broken pipe. The cosmonaut crew that was sent to repair it barely escaped with their lives when their launch rocket exploded. But in 1984, a spectacularly successful eight-month mission had seen the fuel line repaired and some new records set before the crew returned to Earth in November.

So Western observers had no doubt that 1985 would witness further activity aboard the station. Further tests with modular assembly to allow for the creation of larger stations were expected, along with longer flights and "hot handovers" from one space station crew to the next (until then, each expedition had been separated from the previous one by several months). Rumor had it (correctly, as time would show) that the Soviets were preparing for another spectacular "space stunt," involving an all-woman mission to the space station.

What observers (myself among them) didn't know at the time was that *Salyut-7* was completely dead. All radio signals had ceased. Even the Russians at Mission Control in Moscow had no idea what had caused the sudden death of the station. It could have been just a burnt-out radio receiver, or a massive internal fire could have gutted the entire station. The silence of the radio allowed for these and many other possibilities.

There was only one way to find out what had happened, and that was to send up a team of cosmonauts to inspect the station. If the problem was minor, they might even be able to fix it. But the main goal of the flight would be to find out what had gone wrong so that the equipment for the next Soviet space station, *Mir*, could be modified to prevent a recurrence.

Any mission to the ailing station would face a major roadblock, however. Without a radio beacon from *Salyut-7*, the normal space rendezvous plan wouldn't work. Ordinarily, linkup relied on a small radar apparatus in the approaching *Soyuz* that triggered electronic echoes from a corresponding radio on the space station. The timing of these responses was used to measure the range and speed between the two spacecraft, and these values in turn defined the step-by-step series of "braking burns" made by the approaching *Soyuz*.

Without the data from the electronics units (called transponders), the *Soyuz* crew would need new equipment to measure their range and speed. This was especially important during the last few miles, when the two spacecraft were too close to each other to be distinguished by ground radar. To fly this mission, their first-ever rendezvous with a dead target, the Soviets chose 43-year-old Vladimir Dzhanibekov, a four-time space veteran and Russia's "rendezvous ace."

Dzhanibekov was well known and well liked both by his Russian colleagues and by the Americans who had worked with him during the *Apollo-Soyuz* mission 10 years before. By all accounts, he was a skilled pilot and a decent human being. On several occasions, according to American astronaut Vance Brand, Dzhanibekov (the Americans called him "Johnny") privately apologized to the Americans for the strident antireligious propaganda that they were frequently exposed to during tours and visits.

Born in Tashkent, Uzbekistan, during World War II, he attended an air force college, and in 1965 he became a "pilot-engineer." He married an Uzbek woman whose family name had died out, and he exchanged his Russian family name, Krysin, for hers. On space flights, he used the radio call sign "Pamir," a mountain range in Central Asia.

Dzhanibekov became a cosmonaut in 1970, and he commanded four short space missions between 1978 and 1984. Each involved a new and difficult type of space rendezvous. By early 1985, he was back "in rotation" for a new crew assignment. He hoped that it would be the months-long expedition he had always dreamed of.

Dzhanibekov lived with his wife and two young sons in Star City, the cosmonaut village northeast of Moscow, where the Gagarin Cosmonaut Training Center was located. One late winter evening, when the family was discussing weekend plans, senior cosmonaut Aleksey Leonov dropped in without warning. Dzhanibekov knew immediately what the visit meant, since for the past few days he had been helping to plan the *Salyut* rescue mission. "Vladimir, you need to report for flight physicals tomorrow morning," Leonov told him, ruining his plans for the weekend and for months to come. All of his time would be occupied with preparations for his fifth space flight.

For three months, Dzhanibekov trained with his crewmate Viktor Savinykh, a space engineer and veteran of one mission. Savinykh, 45, was a native of Siberia, and he loved cold weather. Soon, he would feel right at home aboard *Salyut-7*.

By the time winter was over in the Soviet Union, the crew was ready to rescue *Salyut-7* from a perpetual space winter. At the Paris Air Show in May, European, Asian, and American aerospace experts swapped stories with one another and with their colleagues from the Soviet Union. A visiting cosmonaut whispered that a new Soviet mission was planned for June, so Western anticipation quietly grew.

Then, with no further warning, on Thursday, June 6, an official announcement from Moscow revealed that *Soyuz T-13* was in orbit, headed for *Salyut-7* for "further work." Officially, all was routine.

Privately, of course, the Russians knew better. Expecting the worst, they had taken additional measures to mentally prepare the cosmonauts. As part of the preflight psychological conditioning, the crew listened to a taped message from a World War II fighter ace: "Have a good trip, my sons, this is Ivan Kozhedub talking to you, who knows what it means to embark upon a combat mission," the tape said. "And that is precisely what you are facing now. I know that you have been given a job of unprecedented difficulty, one which will require all of your abilities, courage, and will. I believe you will do everything to complete your mission. Just like us, pilots of wartime, you have been indoctrinated by our party, the people and all of our Soviet reality."

As he reported to the launch pad for blastoff, Dzhanibekov saluted the officials in charge and vowed, "We will do everything possible." But in hindsight, Feoktistov and others realized that Dzhanibekov had been wrong. The cosmonauts wouldn't just do everything possible; they would do the impossible.

Viewed from the West, strange features appeared almost immediately in the new *Soyuz* mission. For one thing, the cosmonauts were following a slow two-day approach to their target, a unique flight plan. All 40 previous Soviet human-piloted rendezvous missions had used a 24-hour profile. Later on, it would be realized that the approach was chosen to allow for more precise radar tracking and course measurements. I was particularly interested, because orbital rendezvous was my professional specialty for NASA at Mission Control in Houston.

Using navigation data sent from Mission Control in Russia, the *Soyuz* got to within a few miles of the silent station. But then it was on its own. Since the forward-pointing radar antenna was useless because of the station's broken echoing transmitter, the cosmonauts had to measure the angular size of their target and convert the result into distance. The bigger the station looked, the closer it was, and mission planners had calculated exactly how the apparent size of the station related to its proximity. The cosmonauts also kept on zapping the station with a handheld laser range finder, not unlike the ones used by land surveyors in the United States.

To get the best laser view of his target, Dzhanibekov had to approach the station sideways so as to have the *Soyuz*'s biggest porthole facing it. Even though this meant that all of his hand-controller inputs had to be shifted 90 degrees, he skillfully slipped the *Soyuz* right up alongside the *Salyut* while Savinykh, at the window, called out distance and speed estimates. This spectacularly smooth piece of space piloting had never been attempted before (and has not been needed since). But it was just the first of many astonishing accomplishments that Feoktistov was dying to brag about in public, yet wasn't allowed to.

The cosmonauts immediately reported a disturbing feature in the station's appearance: Its two solar-power panels were out of alignment, a sure indicator that the internal computer control system wasn't working. The hope had been that the only failure was in the radio, and that all else on board was normal. But the sight of the station eliminated any such possibility.

Dzhanibekov guided the probe on the nose of the *Soyuz*, which looked much like a harpoon with a latching tip, into the docking cone of the *Salyut*. He then activated motors to retract the probe, pulling the two vehicles together so that the latches around the corresponding tunnel openings could engage the hooks on the station. All of this was standard. It was to be the last "normal" piece of space flying that either cosmonaut would do for many weeks to come.

Through a small valve in the middle of the hatch, they sucked some air out of the *Salyut*. There was no sign of smoke or poisonous fumes. Having eliminated the possibility of a fire, the cosmonauts opened the door and floated in. After a brief survey, they returned to the *Soyuz* to use its radio to report to Mission Control in Moscow.

For years, radio amateurs in Western Europe had been eavesdropping on the conversations of Soviet cosmonauts in space, and they had been publishing their results in such respected periodicals as the British Interplanetary Society's monthly magazine, *Spaceflight*. On this mission, however, they could hear only a few minutes' worth before the spacecraft passed eastward over the horizon and out of range of Western ground sites. The eavesdroppers did hear a few words, however, such as "very cold" and a reference to dust.. Somebody could be heard coughing almost continuously. Then the signal faded.

At Mission Control, flight director and former cosmonaut Valeriy Ryumin was the voice at the other end of the conversation. Dzhanibekov was describing the situation on board the station. "It smells familiar," he said (he'd been aboard twice before). The two men had removed the window blinds, allowing sunlight to flood into the station. Specks of dust floated immobile in midair. The station was dead— Savinykh checked a few electrical outlets and found zero power—and frozen. Dzhanibekov later recalled his overwhelming first impression: "Inside, the situation was even worse than expected."

"It's cold," Dzhanibekov reported from the warmth of the *Soyuz*. "We can't work without gloves. There is frost on the metal surfaces. It smells of stagnant air. The station is in perfect order, but nothing is working." He later expanded on his impressions: "Without gloves your hands get stiff, but with gloves you can't do much. It's like repairing a car in an intersection on a freezing cold night."

One of the cosmonauts' first tasks was to measure exactly how cold the station was. The battery-powered thermometers they had with them went down only to zero degrees Celsius (32°F), and the *Salyut* was clearly a lot colder than that. Savinykh thought of a useful trick from his childhood: Spit on the metal surface and count the seconds until the water froze. They soon figured out that it was −7 or −8°C (around 20°F). That's about the normal temperature of any object as far from the Sun as Earth, but on Earth's surface, the "greenhouse effect" of the atmosphere raises this value significantly.

Meanwhile, the Western public was getting its most direct insight into the mystery mission from the amateur radio listeners, such as Geoffrey Perry of the Kettering Grammar School in Great Britain. Perry also noticed something else that was unusual: A day after docking, the crew was still using the radio in the *Soyuz*, whereas in the past, they had always switched to the *Salyut*'s radio within hours. He put the pieces together. "It really is quite remarkable," he stated. "It fits all the scenarios of a radio command failure."

Before the flight, Soviet experts had decided that if the station's power was found to have failed, the cosmonauts were to remove the central control computer and return with it to Earth. It would have data in its memory that might allow a diagnosis of the failure.

Dzhanibekov and Savinykh were in no hurry to head for home, however, given the trouble they had taken to get on board the station. Mission Control suggested that if they could get behind some of the panels and detach the power lines coming directly from the solar arrays, they might be able to hook them to one of the batteries and charge it up. They could then use its power to bootstrap the rest of the station.

Such an ambitious space repair had only recently become realistic. The year before, both the Americans and the Russians had conducted revolutionary in-flight maintenance on space vehicles. NASA astronauts aboard the *Challenger* had grabbed an ailing science satellite with the shuttle's robot arm and then walked in space to replace broken control circuits and science instruments. A few months later, cosmonauts aboard the *Salyut* had made a series of difficult space walks to open access panels at the aft end of the station, isolate a broken fuel pipe,

install bypass lines, and restore the system to full capability. These successes in 1984 had encouraged even more ambitious attempts the following year.

The cause of the original crisis on *Salyut-7* was quickly determined.
A power-level gauge in the control circuit on one of the batteries had
failed "high," falsely indicating that the battery was fully charged. As a
result, electric current from the solar arrays had been shunted aside.
After several hours, when the battery was almost completely drained,
the control circuit realized that something was wrong and commanded
a power feed to another battery. However, because of one of those
design flaws that is all too obvious in hindsight, the electrical relay
switches that would have conducted this switchover were powered by
the drained battery. The switchover never occurred, the battery went
dead, and with it went the station and all of its electrical equipment.

The repair would be trivial indeed. Once the batteries were fully
charged using manual connections, the faulty control unit could simply
be disconnected.

But this simple repair would have to wait until the consequences of
the failure could be dealt with, and this was a serious challenge. Only
one of the cosmonauts could work aboard the *Salyut* at a time. Without
fans to move the air, their exhaled breath, rich in carbon dioxide, collected around their heads. This bad air led to headaches, rapid heart
rates, sweating, and dizziness. The cosmonauts often had to stop working and fan their faces. Each man took his turn working, then he would
wait in the *Soyuz* (with its heat and its air fans), ready to rescue his buddy
if he lost consciousness.

Over a period of days, the cold seemed to be getting worse.
"Apparently, in zero gravity legs are particularly sensitive to cold,"
Dzhanibekov later recalled. "Fur boots and warm socks were not of
much help. But we quickly found a way to warm ourselves—just tap-
ping one leg against the other while working."

Since the two men kept going back and forth between the frozen
station and the *Soyuz*, and since the hatch had to be kept ajar so that
they could talk and so that the bad air from the *Salyut* could be
processed by the *Soyuz* equipment, the chill of the *Salyut* crept into the
Soyuz as well, despite its electric heaters. "The temperature in the *Soyuz*
came down, and it was pretty cold," Dzhanibekov continued, "because

we had to keep the hatches open and use the ventilation system of the spacecraft."

Savinykh recalled brushing hoarfrost off the station's windows to peer out at Earth. Later he was asked if he had found the cold and ice depressing. "Not at all," he answered cheerfully. "I was raised in Siberia, it reminded me of my childhood, I enjoyed it."

In addition to the cold of deep space, the men were confronted with the absence of any of the mechanical noises normally heard aboard a functioning spacecraft—fans whirring, clocks ticking, and pressure regulators chattering. It was, in Dzhanibekov's words, "the genuine silence of space."

They soon wearied of the silence. The station's tape recorder had malfunctioned. "We need music so badly now," Dzhanibekov reported to Earth after a few days, and it wasn't just for entertainment. Like many pilots and deep sea divers, he began to hear a continuous buzzing in his ears, a constant annoyance that could be relieved only by soft music. Later, a robot supply ship would deliver a new tape player with music, sounds of rain, rustling leaves, birds singing, and other reminders of Earth. But for now, the silence was just one more torment to be endured.

Back on Earth, the handful of observers who knew what was going on marveled at the crew's success. Feoktistov thought of a unique metaphor for the space mission. "Problems, like a mountain, grew in front of the engineers on Earth and the crew," he would write in his report. And they conquered every one.

Two days after docking, on June 10, the crew got the first of eight main batteries charged, and they managed to turn on two small lamps inside the *Salyut*. Two days later, with additional batteries charged, they turned on the *Salyut*'s radio for the first time. On one pass across Europe, they were heard talking on the *Soyuz* radio, and then on the next pass, an hour and a half later, they were using the *Salyut*'s radio (which used a different frequency). According to Geoffrey Perry, listening by radio from England, "They've got a working spacecraft now."

This was not quite true. Although they had power, they still didn't dare to turn on the heaters because the interior equipment was still coated with ice. If that ice were to melt too quickly, the water would soak the electronics, probably ruining the critical systems that had survived the freeze.

Some ice had to be melted soon, however, for drinking water. Aboard the *Soyuz*, they had brought enough to last a week (until June 14). Aboard the *Salyut*, they found two portable water bottles, which they brought back into the *Soyuz*. The bottles thawed in a few days and provided another week's water supply.

The main water tanks on the station were frozen solid, however, and simply turning on the electric heaters would have been disastrous. Never designed for thawing ice, the heaters would have melted a small pocket around themselves, then boiled that water until the entire ice block and tank ruptured explosively. So instead, the heaters were turned on and off, day and night, in a carefully calculated cycle that would slowly melt the ice.

As it turned out, this took only four days. On June 16, the first water ran from the station tanks. "The crisis was over," Feoktistov later wrote.

That same day, with full power restored, the crew did a TV show for Earth viewers. Ever careful to present a false front of total success, Soviet news officials ordered the cosmonauts to remove their coats, hats, and gloves, and smile warmly in front of the camera in shirt sleeves, even though the air temperature was barely above freezing.

One more system remained to be thawed and checked, and that was the *Salyut*'s steering jets. The propellants, nitrogen tetroxide and hydrazine, might also have been affected by the cold, but a careful test firing of the small thrusters showed that they could be used to point the *Salyut* in any desired direction. With this last hurdle cleared, a robot *Progress* supply ship was launched on June 21. It carried replacement equipment, such as batteries and space suits. Two days later it docked automatically, guided by the station's revived radio transponders.

The launching of the robotic *Progress* spacecraft was the final proof to Western observers that behind the Soviets' façade of normalcy, the station had been truly restored to full functionality. On June 22, I circulated a report to a few journalists in which I described what I saw as the unsung triumph of the mission. I claimed that it was largely due to the "boldness and flexibility" of the Soviet space team.

In the view that I expressed at the time, the success of the *Soyuz T-13* in snatching the *Salyut* space station back from the jaws of orbital death would encourage Soviet space planners to bolder space ventures. Although it was less heralded than the dramatic American space shuttle repair and retrieval missions of the same period, the Soviet experience demonstrated equivalent capabilities and taught the same lesson: Complex and powerful space missions are enhanced (and often entirely enabled) by the presence of flexible, skilled human beings in orbit. I felt that such bold successes would naturally reproduce themselves and encourage confidence in future adventures. An entirely new generation of Soviet space vehicles was believed to be under final development, and the Soviet space program was promising to push up and out to a new, higher plateau of capabilities in the very near future. I believed that the saga of the *Soyuz T-13* was still being written, and I predicted that the sagas of its successors would be even grander.

On the basis of my report, buttressed by other interviews, the *New York Times* published an article on June 24, entitled "Soviet Astronauts Rescue Derelict Salyut 7." It was the first news media description of the significance of the repair mission. I called the event a major coup. The story concluded with my assessment that the mission was a very impressive demonstration of Soviet competence and flexibility. I noted that we had always thought of such characteristics as exclusively American.

When the article went to press, it would still be weeks before the cosmonauts fully restored the station. Even after the first robot supply ship arrived, they couldn't switch on the heaters in the hull. If they had done so, the ice there would have melted and evaporated, then immediately recondensed on the still-too-cold electronic equipment. Instead, they had to warm the air inside first, then wait for days while the air warmed the equipment. Then and only then could they thaw the hull.

When they did, it was like a monsoon. "The warmer it gets, the wetter it gets," one of them radioed to Mission Control. "We need a lot of dry rags and a bucket. I'm joking about the bucket."

When he returned to Earth, Dzhanibekov provided more details in interviews. "When the temperature started to rise, the condensation in the station changed to a kind of flood," he explained. "Humidity was rising and it was very hot, but condensation was on the windows and the coldest parts of the equipment and the walls of the station. And it was a long time that we lived in very difficult humidity conditions."

Meanwhile, back on Earth, Feoktistov was continuing his battle with the Soviet news censors. By late July, the station's recovery was

nearly complete. Feoktistov also had the June 23 *New York Times* article to wave in the faces of the censors. It showed them that the Western
reaction to the cosmonauts' accomplishment was one of respect and
praise, not of mockery and denigration. Indeed, that last argument
seems to have been the clincher. On August 5, Feoktistov's paean was
published in *Pravda*. "Courage of the 'Pamirs'" was the title, referring to
the radio call sign of the two cosmonauts.

"It must be said that Vladimir Dzhanibekov and Viktor Savinykh,
having shown true courage under very difficult conditions, carried out
all the operations for reactivating the space station with great care,"
Feoktistov wrote. "The selfless work of the personnel of the Control
Center and all the engineers who took part in analyzing the situation
and developing the programs and methods to restore the space station's
capacity to operate, cannot be left unmentioned."

"There has not been such a difficult flight in the practice of Soviet
cosmonautics," cosmonaut corps commander Vladimir Shatalov was
finally allowed to say in public. "We have never before encountered such
difficulties."

Dzhanibekov stayed aboard the station for 108 days, making a
space walk and welcoming the next crew. He thus became the first man
to hand over the command of a space station to a successor. After landing, he took a job training cosmonauts, barely survived a terrible auto
accident, and got on with his life.

The cosmonauts were not the only ones to achieve great things in
the mission. Among the Russian space engineers on Feoktistov's team
who worked nonstop for months to make the rescue mission possible
was an activities scheduler named Sergey Krikalyov, then 28. His performance was so impressive that his application for cosmonaut training
was accepted within a year. Soon afterward, in 1988, he flew into space,
becoming the first-ever "citizen of the Space Age" to do so: He was the
first human space traveler to be born after the launch of *Sputnik*.

But the successes would not last forever. All that was achieved by
the ingenuity and endurance of Dzhanibekov and Savinykh aboard
Salyut-7 and by the dogged, determined support of their ground crews
back on Earth was canceled out by a setback that put an end to most of
the plans for the station. The crew that took over for Dzhanibekov and
Savinykh succumbed to medical problems that couldn't be treated

with the frost-damaged contents of the spacecraft's pharmacy, and the cosmonauts had to evacuate the station. A microcosm of the Soviet space effort as a whole, *Salyut-7* ended up falling prey to a large accumulation of negative factors.

In order to understand the origin of the near-fatal influences that were threatening the entire Soviet space program, it's necessary to look a little further into the past. As it turns out, these factors were also the basis for the plans that were made in the late 1990s, and they directly affect the events that shape the direction of space exploration today.

2

History

One of my most cherished space artifacts is a lump of aluminum. About the size of my palm, it shows the marks of the gouges and blows it suffered when it was still half molten in midair. The technobabble phrase "severe thermal and mechanical stress" comes to mind when you see it.

It's a fragment from a dream that died, a piece of the doomed rocket Russians built to send humans to the Moon. Growing up in the Space Age, I shared in that dream, eventually becoming an engineer at Mission Control in Houston. But for many of the dreamers, the dream had turned into a nightmare.

I once had the opportunity to console some of those tortured dreamers. Back before NASA's space partnership with the Russians had been formally established, Russian visitors to NASA centers were scrupulously cold-shouldered. But since I was widely known in Houston as the "Soviet space buff," I was often telephoned at my desk at the Johnson Space Center and asked to stand in for officials who didn't want to be seen meeting with unapproved foreigners.

As NASA's entirely unofficial "ambassador to the Russian space program," I once brought a group of rocket engineers from Dnepropetrovsk to a barbecue for friends and fellow space workers. It was 1989, and the Soviet group was touring the United States, trying to sell its *Tsyklon* and *Zenit* rockets to American clients. The leader was Yuriy Smetanin, a wiry old Cossack who grew more cheerful the more he drank. In a fit of sudden bravado, he leaped onto one of the kids' saddled horses and galloped off into the woods behind the house. Fears of an international crisis faded only after he returned, safe and sound and singing boisterously.

Other fears lingered, however. At least a dozen NASA workers at the party privately asked me if I thought that they should formally report their unofficial contact with "Soviet agents" when they returned to the office on Monday. NASA's attitude toward quarantining Soviet space experts was deeply ingrained. So, too, I was to learn, was its dismissive attitude toward Soviet space expertise, an asset that NASA seemed unwilling to seek out and utilize.

So it's not surprising that officials were in a quandary when the head of Russia's human space program asked for a tour of NASA's Johnson Space Center, south of Houston. Yuriy Semyonov was the head of a delegation of Russian space workers who were visiting Houston to attend a local space conference. But although the local protocol officer met them at the center and even found a Russian emigré to interpret, no astronaut or manager dared to show up for the tour—they feared the political and professional ramifications of looking "too friendly." So the protocol officer telephoned me and asked me if I'd like to come along. I was happy to, and after formally requesting a day of leave so that the meeting would not be job related, I went down to the headquarters building. I knew the drill.

Semyonov had just taken over the leadership of what was now called the Energia Rocket and Space Corporation, or RSC-Energia. This was the complex of laboratories, design bureaus, and fabrication halls that built and operated Russia's space vehicles with cosmonauts on board.

RSC-Energia was only the latest incarnation of a space team that had originally been founded in 1947 by Sergey Korolyov, the legendary "Father of the Soviet Space Program." Then called OKB-1, or "Design Bureau 1," it was responsible for building most of the USSR's early missiles. This was the team that had fired the opening shots in the Space Race, launching *Sputnik, Vostok, Lunik,* and other "firsts."

While still in his thirties, Semyonov had been in charge of the Soviet program to send two cosmonauts out around the Moon and back, ahead of *Apollo.* Other engineers were working on the even larger rockets and spacecraft that would be needed for a lunar landing by cosmonauts.

But Korolyov had died unexpectedly in 1966, his heart weakened by years in Stalin's labor camps. He was only 59 when he died, and he had been the chief organizer of the USSR's space and missile industry for almost 20 years. The technical challenges mounted, and NASA's *Apollo* program picked up steam. Without a leader, the Soviet effort stalled, then fell apart in a series of secret disasters. Perhaps the challenges would have been too great even if Korolyov had lived, but without him, the Soviets were simply devastated.

So here I was, face to face with one of the losers of the "Moon race," one of Korolyov's chief lieutenants on that project. He had gone on to lead more successful efforts, helping to develop the series of small inhabited space stations called *Salyut,* then working on the Soviet answer to NASA's space shuttle, the *Buran.* We stopped the minibus by the *Saturn-V* Moon rocket, which was lying horizontally on the lawn in front of the Johnson Space Center. Semyonov and his entourage walked around the vehicle that but for budget constraints would have carried three astronauts to the Moon. While fascinated with the hardware, the Russians grew glum and introspective as they recalled the years that they had devoted to the Soviet equivalent. "We did not succeed," one of them remarked, "and our failure was a bitterness which has not softened over the years."

Attempting to console them, I stretched my conversational Russian—and my repertoire of hand gestures—to the limit. The Soviet program was not a waste of their lives, I insisted. I pointed out that the plaque that was left on the Moon merely says "men from Earth," and I handed out some small memorial pins that represented the actual plaque. I told them that the astronauts went to the Moon "for all mankind."

They already knew all the words, and the words were no consolation. The pins, however, were a modest hit. The Russians fumbled in their pockets for some of their *znachki* pins, small plastic emblems, to return the favor.

I tried a different approach. I told them about how I remembered seeing the spy satellite photographs of the Russian Moon rocket as it was being prepared for its maiden flight. Actually, this was artistic license, since I hadn't actually seen the photographs until the early 1970s, when I was doing technical intelligence work for the U.S. Air Force. The view in the photographs was from nearly overhead, and the rocket stuck up out of the flat Central Asian steppe like the proverbial sore thumb.

I told them about how the United States had thought about quitting after the *Apollo* fire killed three astronauts in 1967 and the Democrats lost the White House in 1968. The threat of the Soviet program was what kept us going, and their rocket was a thumb up our butt, making us jump all the way to the Moon. I used some graphic hand gestures to underscore the final point.

I told them, sincerely if a little theatrically, that I doubted that the United States would have had the willpower to complete the *Apollo* program had it not been for the advanced state of the Soviet program. I assured them that their effort had not been for nothing, that it had actually made the Moon landing possible.

Semyonov and his associates paused to ponder this new interpretation. Or maybe they paused to wonder about my awkward Russian circumlocutions. Then, one by one, they began to nod and smile. One reached over and gave me a big hug.

My reasoning was only a short-term emotional Band-Aid, however, a spoonful of chicken soup for the Russian space soul. Because for an entire generation of Soviet rocket scientists who had devoted their lives to opening the space frontier for the entire world, the triumphs seemed to always be overshadowed by the tragedies. And there was nothing that anybody could ever do about that.

In 1991, Russia's top space historian, Yaroslav Golovanov, gave a belated obituary for the losing side of the Moon race. "The unfulfilled dream of Sergey Korolyov, who died on the operating table—a dream that was decimated by Valentin Glushko, that was undefended by Vasiliy Mishin, and that took years of labor by Nikolay Kuznetsov—vanished in the gulf of ministerial paperwork and the flames of failed launches that turned billions of rubles into ashes," he wrote.

These unfamiliar Russian names—Glushko, Mishin, and Kuznetsov—could just as well have been characters from *War and Peace* as far as the Western space experts of the 1960s were concerned. Under the Soviet regime, all of the leading Soviet space officials were kept strictly anonymous. Behind the façade of communist-led centralized efficiency lay bitter rivalries, even enmities, that bedeviled Korolyov's drive to organize a winning space team and, after his death, scuttled the loyal efforts of his successor, Vasiliy Mishin.

About eight years after Korolyov's death, Mishin vanished into a "memory hole," exiled to a teaching job and forbidden to speak in public. He was kept silent for almost 20 years as a result of a bureaucratic coup led by Valentin Glushko. Glushko had once been one of Korolyov's most talented lieutenants, but in later years, he tried to rewrite history to place himself at the center of all Soviet space successes. Under the relaxation that took place as the Soviet state decayed, however, and especially following Glushko's death in 1989, Mishin was able to emerge from the shadows and describe his own role, as well as his feelings about the loss of the Moon race.

"We felt a deep sense of sadness," he wrote in 1990. "It was a colossal project to which we had dedicated our best years. I was young at the time. And it was the work of a great many people and it vanished overnight. The Americans had won. I was made the scapegoat."

I was a witness to the gradual process by which the real history of those years oozed out through the wilting barriers of the dying Soviet empire. Here and there, I was able to help this process of discovery along. It was exciting, frustrating, and often misleading.

In 1989, for example, an old Russian rocket scientist leaned over to me and whispered the secret name of Korolyov's Moon rocket into my ear. The rocket had been designed to carry cosmonauts to the Moon ahead of the *Apollo* astronauts. My source was more than a little drunk, sitting with me in a noisy Moscow restaurant, and I was more than a little confused by the unfamiliar jargon. So as he enunciated the three-syllable name, I thought I heard him say "en-er-din." I assumed this was a combination of *energia* (energy) and *dinamika* (dynamics). Only months later did I realize (thanks to some friends with better hearing and more clear-tongued sources) that what he'd really been saying was "en-uh-din"—that is, in Russian, *N odin*, or *N-1*. The "N" was for *nosityel*, Russian for "booster," and the *N-1* was the biggest rocket that humans have ever built, then, now, and perhaps for centuries to come. Even bigger than the rocket, however, was the veil of secrecy that had been cast over its existence, its history, and even its name.

N-1 rockets were launched four times between 1969 and 1972, and they all failed. In hindsight, the main reason appears to have been a falling out between Korolyov and Glushko, who had built the rocket engines for all of Korolyov's previous space launch vehicles.

Korolyov had favored the development of high-powered but hard-to-handle cryogenic (very low temperature) fuels such as liquid hydrogen, which NASA had also chosen for its *Saturn-V*. Glushko, however, insisted on sticking to the traditional room-temperature self-igniting fuels and high-efficiency engines that he had been improving year after year. When the decision went against him on the *N-1* rocket, Glushko refused to cooperate, and Korolyov turned to an aircraft engine designer named Nikolay Kuznetsov to build the rocket engines that he would need. Indeed, Kuznetsov's efforts were awesome, and he almost succeeded.

Meanwhile, off to the side of the struggle between Korolyov and Glushko, another space personality was playing a powerful role. This was Vladimir Chelomey, an aircraft designer who later graduated to cruise missiles, space launch vehicles, and killer satellites. A decade younger than his rivals, his ultimate goal was to entirely supplant the bickering Korolyov/Glushko camps and take leadership of the entire Soviet space program himself.

This is not ancient history, and these secret Russian space rivalries aren't akin to some obscure dynastic struggle in sixteenth-century China. As astonishing as it may seem today, at the dawn of the twenty-first century, when all the principal actors are long dead and their entire country has gone through revolutionary upheavals, the back-and-forth battles of those years left scars that are still visible. The legacy of the old Soviet agencies, institutes, and alliances still defines the underlying relationships within today's Russian space industry, and the very design of the *International Space Station* now in orbit reflects the personal animosities of the Russian space figures of the 1960s.

The very first Soviet space station, code-named *Zarya*, was built through the forced partnership of two rival teams, the Korolyov group and the Chelomey group. In the 1960s, Chelomey was designing space stations for cosmonauts, while Korolyov was concentrating on flight to the Moon. At first, Chelomey's growing influence seemed certain to overtake Korolyov's, especially since he had hired Nikita Khrushchev's son Sergey as a project engineer. Sergey was actually a talented rocket scientist in his own right, but his greatest asset for Chelomey was clearly the "in" that he provided with his father.

What's more, the Korolyov-Chelomey rivalry reflected a bureaucratic battle that in turn reflected a critical cleavage in Russia's space industry. From as far back as the late 1940s, Korolyov's efforts had been supported by the Soviet artillery industry, whose primary interest was in throwing warheads as far as possible. Chelomey, on the other hand, represented the aviation industry, which had pushed for the winged cruise missiles that competed with Korolyov's rockets as intercontinental weapons. Each side had its supporters and detractors high within the Soviet government. And despite Korolyov's record of success in the early Space Age, Chelomey's "insider" maneuvers gave him an advantage.

Impressed with Chelomey's plans, Nikita Khrushchev transferred a number of space and missile facilities to his empire in 1960. One was a missile design institute in a south-central Moscow neighborhood called Fili. An automobile plant that had begun building airplanes in 1923, the complex had by 1951 become a major bomber plant. Under the directorship of the brilliant aeronautical engineer Vladimir Myasishchev, a specialized team had been established to design advanced aircraft. In the late 1950s, the group began working on missiles from its offices next to the aircraft factory. It was this group, which later became known as the Fili branch, that was assigned to work on spacecraft for Chelomey.

Then, in 1964, Khrushchev was ousted. With the advantage of family connections gone, Chelomey's influence declined, although he was still supported by many military officials, among them Andrey Grechko, who would later become the Soviet defense minister. Although Chelomey's proposals for lunar spacecraft and super-boosters were rejected in favor of Korolyov's, he was allowed to complete the development of the heavy military missile that eventually became the *Proton* space booster, now the main heavy booster for Russian space launches. Another project he continued was the *Almaz* orbiting military reconnaissance outpost, as well as its special piloted supply ship, the *TKS*.

What happened next is concisely described in historical reports at the *Encyclopedia Astronautica* Internet site. One such report succinctly describes the time: "After America's landing on the moon and the explosion of the second *N1* [July 1969], the engineering staff at Korolyov's bureau found itself with no immediate work at hand. There would be a long delay for redesign of the *N1*, and they had spun off all of their unmanned spacecraft to other design bureaus."

"The next round of the space race would be the first manned space station," the article continues. "NASA's *Skylab* was scheduled for launch by 1972. Like-minded engineers conspired behind the backs of Chief Designers Chelomey and Mishin. Through Communist Party channels they proposed to take the *Almaz* hulls already built by Chelomey, outfit them with flight-qualified *Soyuz* systems [from Korolyov's bureau], and launch the spacecraft before *Skylab*. The Soviet leadership, fed up with the internecine quarrelling that contributed to the loss of the moon race, agreed. The *Salyut-1* long duration orbital station design was created."

Prime Minister Leonid Brezhnev himself intervened to make that decision, and he would intervene repeatedly in the coming years. It was an unstable situation, with spacecraft from two antagonistic groups—Chelomey's and Mishin's—being manufactured in different halls of the same factory. Obsessed with the human lunar mission, Mishin had never wanted to involve his team in the space station project. His lack of support for it led to several attempts to remove him from control of the "Korolyov team."

Shortly before its launch, although the original name *Zarya* was still painted on its side, the station was renamed *Salyut* in honor of the tenth anniversary of Gagarin's flight. But the mission was a replay of the Moon race disaster. A mechanical failure in the docking mechanism prevented the first crew from boarding the station. It even almost prevented them from undocking, which would have proved fatal. Although the second crew did manage to board the station and work there for 21 days, the station's main scientific instrument, a solar telescope, was useless as a result of its failure to jettison its protective cover. During their return to Earth, those second crew members were killed when their *Soyuz* spacecraft accidentally depressurized. After redesigns, a second space station was launched in July 1972, but its rocket blew up soon after liftoff.

As a result of these failures, the two antagonistic space teams split up and returned to their own separate projects. Now each group was ordered to prepare a space station that would fly before America's *Skylab*, which had been delayed until May 1973. Brezhnev selected the Chelomey design to fly next, and it was launched in April 1973. Shortly before its launch, however, the Korolyov group (which had built the *Soyuz* that was supposed to carry the cosmonauts to the station) announced that they were grounding the *Soyuz* for several months because of a recently discovered parachute problem. The defect was probably real, but the Chelomey team can be forgiven some dark suspicions of deliberate sabotage.

Since the Chelomey vehicle had already been loaded with propellants, and since recycling the space station for launch at a later date would have taken an exhaustive rebuilding process, the station was launched anyway in a purely automated test. Just 11 days after launch, however, on-board systems suddenly reported that the internal air pressure was dropping. The station died within a day and then began to tumble. It burned up in the atmosphere on May 28.

For years, engineers tried to explain the failure. First they blamed the propulsion system, then they considered an electrical fire. They also wondered if the station had collided with a fragment of its booster rocket, which had exploded in a nearby orbit. Nobody could figure it out. But the opinion persisted that if a crew had been launched to the station in time, they could have diagnosed and repaired the flaw before the station was lost.

The Korolyov team now had its turn at bat, and it launched a space station into orbit just three days before *Skylab*. Perhaps the team had suddenly found a way to requalify the *Soyuz* ferry ship for carrying cosmonauts to the station once it was in orbit. However, a software flaw in the station's computerized autopilot caused it to expend most of its stabilization fuel while it was out of radio contact with Russian ground sites. As a result, the station had to be abandoned. It died so quickly that it wasn't even officially called a *Salyut*; instead, it was given the deceptive name *Kosmos-557* to disguise it as a routine research satellite. Since at least a year and a half would be needed to redesign the project, the cosmonaut crew was reassigned to a different mission. The next station from the Korolyov team, *Salyut-4*, flew successfully in 1975.

By then, the Chelomey team had succeeded in flying one of their stations, the *Salyut-3*. The most bizarre feature of this vehicle wasn't its spy cameras, which didn't work very well, but its short-range defense mechanism against a feared American attack. The Soviets had installed a modified jet fighter cannon on the nose of the station, and they actually test-fired it in space. The presence of this machine gun (designed by the famous Nudelmann Bureau of aircraft weaponry) on an inhabited spacecraft was one of the most closely guarded secrets of the Soviet space program. It didn't come out completely until the late 1990s.

A second Chelomey space station was launched in 1976 and called *Salyut-5*. While it was a technical success, it failed to demonstrate that cosmonauts provide many additional advantages over automated (no crews) military reconnaissance space missions, and the Soviet Defense Ministry decided to expand its automated space spy satellite program instead. Undiscouraged, Chelomey laid plans for bigger and better stations, both with and without cosmonaut crews.

Competition between the rival space teams remained intense. Yet even as his bureau was preparing its space stations, Mishin hadn't stopped shooting for the Moon. Two more attempts, in 1971 and 1972, to fly the *N-1* booster had failed. Undeterred, Mishin prepared a fifth test vehicle, but the patience of the Soviet government—and of most of his own subordinates—had run out. While he was in the hospital for a routine checkup, he was told that he had been fired and should turn in his badges. His replacement was Korolyov's old rival, Glushko. Glushko's first step after canceling the latest *N-1* rocket was to rename Korolyov's group "NPO Energia" (NPO is Russian for Scientific Production Organization).

With all hopes for making a successful *N-1* abandoned, the giant rocket acquired a nickname, one that reflected the wry Russian sense of humor. The *N-1* became widely known as the "Tsar Rocket." This was not a term of respect; it was a historical allusion to all the previous "biggest-in-the-world" Russian projects that had ended in disaster. As a Russian engineer explained it to me, many years later: "Remember the 'tsar-cannon,' biggest in the world? It couldn't shoot. And remember the 'tsar-bell,' biggest in the world? It couldn't ring. Then we built the tsar-rocket, and—you know what happened."

In spite of Chelomey's recent setbacks, competition for government funds remained strong. In his new position as the head of Korolyov's old team, Glushko found that he still had a formidable competitor for space project funding. Chelomey, backed by Minister of Defense Grechko, pushed for his military-focused *Almaz* space stations. But Glushko accelerated the development of the multimodular research stations that would lead to the *Mir* space station. His backers in the government included top party official Dmitriy Ustinov and Politburo member Andrey Kirilenko, a former aircraft engineer. One of Glushko's allies within the Korolyov group was Yuriy Semyonov, then about 40 and married to Kirilenko's daughter. Semyonov thus became Glushko's first deputy and the space station's chief designer.

Early in 1976, Grechko died (at the age of 73), and Ustinov (then 68) became defense minister. Glushko's triumph was now complete, as Ustinov ordered Chelomey's cosmonaut-related space projects terminated. Glushko, by contrast, was chosen as a member of the twenty-fifth Communist Party Congress.

"From this moment onward, Glushko concentrated in his hands not only the power of an enormous space empire, but also the political power of a commissar, capable of overwhelming anyone in the space establishment," wrote Roald Sagdeyev, a leading Soviet space scientist who later emigrated to the United States. A few years later, Glushko took Chelomey's space institute, the Fili branch, and assigned it to his own organization.

Down but not out, Chelomey still hoped to outlive his older rivals. From his home factory in Reutov, near Moscow, he continued to work on naval cruise missiles, his original specialty. In December 1984, at the age of 70, he was injured in a car accident. While recovering in the hospital, he was cheered to learn that Ustinov had suffered a major heart attack and was paralyzed. Perhaps Chelomey began planning new space projects even then, but just a few days later, while still in the hospital, he suffered a fatal stroke.

The myth of the monolithic "Soviet space program" is shattered by these sad episodes. Yet the myth remained so convenient for Soviet and

American officials alike that it endured and continues to bedevil us even today. By the late 1990s, Americans would be debating about whether or not Russia was a good "partner" on the *International Space Station*. But with no knowledge of Russia's internal pattern of conflict and competition, any such assessment would always fall short.

Another classic case of misperception dominating sound judgment is the *Buran* space shuttle, which flew just once in an automated mode in late 1988 and then was scrapped as a useless detour, after the expenditure of about $10 billion. The project had been proposed in the mid-1970s as a response to fears that NASA was building its own space shuttle so that the U.S. Air Force could bomb Moscow from space.

According to Russian space historian Yaroslav Golovanov, these warnings from military officers in the Space Units impressed neither the Soviet Ministry of Defense nor the defense minister, Grechko. Perhaps his staff correctly advised him that these fears were implausible at best. Perhaps he was just instinctively opposed to expensive space vehicles. He did not buy—and did not pay for—the recommendation.

The campaign then bypassed the Soviet military establishment and enlisted the support of Leonid Smirnov, a senior civilian minister in the aerospace industrial complex. At the end of a regular briefing to Leonid Brezhnev on space program status, Smirnov inserted a brief comment that had not been on the agenda, about the NASA program.

"The Americans are intensively working on a winged space vehicle," he told Brezhnev, according to Boris Olesyuk, a *Buran* manager who read the minutes of the meeting. "Such a vehicle is like an aircraft," Smirnov continued. "It is capable, through a side maneuver, of changing its orbit." In other words, it could suddenly appear over Moscow and drop anything it wanted.

"The news disturbed Leonid Ilyich [Brezhnev] very much," Olesyuk later recalled. "He contemplated it intensively, and then said: 'We are not country bumpkins here. Let us make an effort and find the money.' Of course, nobody dared to contradict Number One."

"Thus . . . an erroneous decision was made by the General Secretary," Golovanov wrote in 1991. "Made by one man with immense power and shallow mind, and on top of that not possessing even elementary knowledge of space issues." Historians have noted that such decisions were common from the aging Brezhnev in those years. As another example, he revived the controversial Stalin-era plan for a new Siberian railway, the Baykal-Amur Mainline (or BAM). The railway would run parallel to the Trans-Siberian Railway, only hundreds of miles farther north, safely away from the Chinese border. He designated it as a "shock project," which meant that it became a national priority, with precedence over other projects in funding, equipment, and personnel. Despite enormous costs and deadly accidents, work on the BAM continued year after year. By 1989, when the work finally ground to a halt, even conservative estimates placed the cost at $13 billion, and the total was probably twice as much when contributing costs were added. The railway never paid off, and even today, it costs more than $200 million in government subsidies to operate.

"Unlike the West, it was not money that guaranteed success in the Soviet Union but rather political will," wrote Athol Yates in the *Siberian BAM Railway Guidebook.* "If a problem occurred, then more labor and resources were simply channeled into overcoming it," he added.

The decree from the CPSU (Communist Party of the Soviet Union) Central Committee and the USSR Council of Ministers concerning the creation of the reusable *Buran* space system was signed on February 17, 1976.

For the next dozen years, apparently nobody in the Kremlin gave the project another thought. Its successful flight test in 1988 was celebrated by Soviet officials, but there was a strong countercurrent in the press. The USSR of 1989 was a very different society from that of previous decades, when space flight was a source of unanimous pride. In a time of economic decline and an easing of international tensions, many experts immediately questioned the value of the project.

In the spring of 1989, a space correspondent from the daily newspaper *Socialist Industry* asked Boris Gubanov, the shuttle's chief designer, "Why do we need *Buran?* The goals achieved by this truly marvelous machine are still unseen. Aren't we being dragged into a senseless and wasteful race?" The chief designer was not allowed to make a coherent reply, except to admit that the spaceship was a bit ahead of its time. The project was stalled, but it lingered near death for several more years before it was officially killed by the post-Soviet government in Moscow.

To see how the old patterns endured even after the deaths of the original protagonists, let's look at one last example. One more space station was to be born out of the forced marriage between the two antagonistic space teams. In the mid-1980s, the Korolyov group (now called NPO Energia) was building a superrocket called *Energiya* to carry its *Buran* shuttle into orbit. The Chelomey group was working on space-based weapons, and they had a payload called *Polyus* that could be launched only by a rocket as large as *Energiya*.

Few details about what was called the *Polyus* satellite have ever been published, but there were reports that it carried an experimental anti-satellite laser weapon called *Skif*. The Russian word might refer to a small boat, but it was also the name of a warlike tribe in Ukraine 2000 years ago, the Scythians.

Soviet premier Mikhail Gorbachev had himself traveled to the Baykonur Cosmodrome in May 1987 to witness the launching of *Polyus*. According to rumor, however, he had been promised only the test flight of a powerful new rocket. In office only a year, he had never been informed of the military nature of the payload. *Polyus* was exactly the kind of space battleship that he had denounced the Americans for proposing.

At Baykonur, he was finally informed about the planned testing of the space weapon. Reportedly, he was furious. His anti-Reagan propaganda campaign would be totally neutralized by the launch of a Soviet rocket carrying the same kind of weapon. So, although he allowed the booster's test flight to go ahead, he also ordered that the payload not be activated once it was in space.

Although the launching of the *Energiya* rocket was successful, the payload failed to reach a stable orbit as a result of a problem with its propulsion system. It was destroyed an hour later when it fell back to Earth over the southeastern Pacific Ocean. Perhaps this was merely an unfortunate accident, but it's possible that the failure was in tacit obedience to Gorbachev's orders.

This mystery mission remains an open question in Soviet space history, but it also has connections to the U.S.-Russian partnership on the *International Space Station*. The propulsion unit attached to the *Polyus*, the one whose failure led to the loss of that hundred-ton space station, was a modified space vehicle used earlier in the military *Salyut* space station program. Years later, its design was improved and a new version was built as part of the *International Space Station*.

That module was called the *FGB*, which in Russian stands for "Functional Cargo Block." Paid for by American money but manufactured in a Russian factory, it was designed to be the cornerstone of the *International Space Station*. It was launched in November 1998 and remains a part of the *ISS*.

I had presumed that in preparation for the *ISS*, NASA experts would have wanted to know what went wrong in May 1987, to make sure that the problem had been fixed before the *Polyus's* sister spacecraft was launched as part of the *ISS*. But when I checked, I was dismayed to find that NASA officials denied that the 1987 failure had ever occurred.

Both in public and in its own internal reference documents, NASA claimed that all previous flights of *FGB*-type vehicles had been successful. In a letter to the *Washington Times* critical of one of my articles, NASA "Russian expert" Tom Cremins wrote haughtily, "In the interest of improving your paper's accuracy, the [FGB] has flown a number of times—all successfully." The third edition of a NASA study, "Soviet Space Stations as Analogs," also described the 1987 mission as successful. This ignorance was apparently based on official Russian assertions to NASA, assertions that NASA had no desire (or no capability) to question. But that self-imposed ignorance would become a theme of the 1990s, and in the 1980s, despite our best efforts, we were all ignorant of the inner workings of the Soviet space industry.

In the mid-1980s, Glushko's team had been neglecting its developmental work on the advanced space stations, concentrating instead on the construction of a space shuttle analogous to NASA's. The space station *Salyut-6* (1977–1981) had been a success, and visits to *Salyut-7* (launched in 1982) were proceeding well, but they represented only slight improvements over the designs from the mid-1970s.

Each of the *Salyut* stations weighed about 20 tons and was the size of a school bus. They had docking ports at each end. While remaining in orbit for several years, they could be visited by teams of cosmonauts using the smaller *Soyuz* vehicles. The greater the number of cosmonaut visits, the more supplies had to be brought up to the station. So during this period, NPO Energia engineers introduced a robot version of the *Soyuz* called *Progress*. All cosmonaut-related hardware, such as life support and manual control equipment, was removed and replaced with badly needed supplies for transfer to the space station.

As they contemplated the succession of Soviet space stations throughout the early 1980s, Soviet space officials expected that the next step, toward the end of the 1980s, would be a modified *Salyut*. It would be called *Mir*, and it would have additional attachment points for large, permanent add-on modules. These would provide specialized laboratory and support facilities.

Early in 1984, Glushko was ordered to prepare *Mir* for launch prior to the 27th Communist Party Congress, scheduled for early 1986. The order came from Oleg Baklanov, the newly appointed minister of general machine building, the Soviet government agency that administered space and missile activities. Baklanov's motivations still aren't clear, but Brezhnev seems to have gotten bored with the repetitive *Salyut* missions. Perhaps he wanted some new "space spectacular" to overshadow Reagan's recent call for NASA to build its own space station. Breakdowns in *Salyut-7*'s propulsion system in late 1983 may also have convinced Baklanov that the station was unreliable.

"While *Mir*'s early 1986 launch may have made sense from a political standpoint," wrote Belgian space historian Bart Hendrickx recently, "one could argue if this timing had any real operational value. The station was underequipped and the first add-on module was still months away." Aside from activating the station, the first crew had little to do that was of any practical value. According to Hendrickx, "Workers at the cosmodrome privately referred to the *Mir* launch as an 'IBD,' Russian short for 'Imitation of Stormy Activity,' which was cosmodrome jargon for a ceremonial space shot."

Satellites circle the Earth along paths that lie in a flat plane that intersects the center of Earth. The orbital plane is usually tilted up to some degree from Earth's equator. *Mir* was originally intended to fly in an orbit inclined by 65 degrees to Earth's equator, rather than the 52-degree inclination of all of the previous Soviet space stations. This steeper orbit would allow it to fly over much more of Russia and the rest of the world. It would also allow rockets from Russia's Plesetsk cosmodrome north of Moscow to enter the same orbit and link up to the station. Stations in the old, lower-inclination orbit had never been able to reach far enough north to pass over the Plesetsk cosmodrome.

The orbit that can be reached most efficiently from any given launch site is the one reached by flying due eastward. To reach orbits with higher inclinations, a rocket must be launched to the left or to the right of this optimal path. But the sharper "left turn" that *Mir* would have needed in order to reach the 65-degree orbit sacrificed too much of the beneficial push from Earth's eastward rotation. Engineers were forced to lighten the payload on *Mir* in order to make room for the extra rocket fuel that was needed to make up the speed. So as *Mir* grew too heavy for its booster to perform the turn, the orbital path was shifted back to the standard 52 degrees.

This minor alteration in the flight plan was to have profound and unforeseen consequences in future years. By making it impossible for space flights from Plesetsk to reach *Mir*, it made the station utterly dependent on the original *Mir* launch site, Baykonur. But Baykonur was soon to be located in a foreign country, Kazakhstan, following the collapse of the USSR in 1991. On the other hand, the more southerly orbit was within the range that American space shuttles could reach (their maximum inclination was 62 degrees, set by the launch path from the Florida coastline). The diplomatic implications of this minor orbital course change would set the tone for Russian-Kazakh relations, and for Russian-American relations, throughout the 1990s.

Mir was launched on February 20, 1986, beginning its illustrious 15-year career. Though it got off to a slow start spawned by old-style Soviet politics, it eventually picked up steam and hosted many spectacular achievements over the years. But that's a subject for later chapters.

In January 1989, Glushko died at the age of 80. He had maintained control of the Soviet space program for 15 years, and he had done his best to impose his will on both its future and its past. He had succeeded in creating a Soviet "space shuttle," the *Buran*, which the government would throw away much too soon, and a multimodular, permanently occupied space station, which the government would eventually avoid throwing away for too long. He failed to interest the Soviet government in his more grandiose plans for Moon bases and Mars expeditions, and he failed in what seems to have been his most passionate quest: to rewrite Soviet space history with himself as the primary figure.

The forced marriage of the Fili branch and Energia was dissolved. In 1993, the Fili branch resumed its original association with the neighboring missile factory that had always been owned by the Chelomey team, now called the Khrunichev Machine Building Plant. Its space vehicles include all *Proton* carrier rockets, the hulls of all of the space stations from *Salyut* to *Mir* to the *ISS Service Module*, and the free-flying specialized modules which docked to *Mir*. Semyonov remained in charge of the Korolyov team, now called RSC-Energia, until he retired in 2001.

Smetanin, the wild Cossack rocket sales representative at our barbecue, clinched the deal he had been seeking. Working with Boeing, his company developed a system to launch his rocket from an ocean-going platform in the Pacific. The first Sea Launch mission, on March 27, 1999, was a perfect flight, and Severin was on the command ship to watch. I never met him again. On his return to Moscow from viewing the launch, he collapsed and died, at age 75, from a heart attack. A veteran of the first phase of the Space Race, he had led the way into a new phase of space partnership, and he had lived just long enough to see it born in flames.

3
Decaying Orbit

"The ultimate result of shielding men from the effects of folly is to fill the world with fools."
Herbert Spencer

"Truth is immortal; error is mortal."
Mary Baker Eddy, *Science and Health*

The most reliable barometer of the health of the Russian space program may be the Kosmos Pavilion exhibit hall in Moscow. For three decades, the condition of the items on display at the exhibit and the mood of the crowds milling past them have accurately reflected the Russian public's changing attitude toward space activities.

When I first visited the Kosmos Pavilion as a graduate student in 1968, it was one of several dozen buildings in a major theme park boasting of the successes of socialism. The Space Race was in full swing, and the USSR appeared to be winning. At the space exhibit, the space hardware was shiny, the paintings of the heroes were brightly lit, and the faces of the many visitors glowed with smiles. In the eyes of the viewers, spectacular space activity symbolized future prosperity. This was their future, worth many sacrifices, and they knew that the world envied them.

In 1989, I hiked through early spring slush and late-afternoon gloom to revisit the same shrine. This time, it was evident that the neglect and stagnation of the Brezhnev era had been overwhelming. The long, dingy hall had the aura of a forgotten tomb. The rotunda area had been cleared and cordoned off because the roof was leaking and parts of the ceiling had fallen on visitors. The windows were dirty, and the lights were dim. Cynicism and apathy seemed to be the dominant public moods, except among those whose feelings toward the expensive space projects that had never had any practical benefit were really hostile. Through a double door, a small side hall offered a colorful, brightly lit temporary exhibit on flying saucers and ESP. Grim-faced Soviet tourists from out of town broke into blissful grins as they entered the world of the paranormal, leaving the disappointments of reality behind.

By 1995, the main exhibits in the Kosmos Pavilion were not about space activities at all. The entire park, once called the Exhibit of Economic Achievements, had become the All-Russian Exhibit Center. It was now devoted to commercial products. The Kosmos Pavilion was full of automobiles and sailboats on display for potential buyers, and the space hardware had been shoved over to the side of the hall or into the smaller, out-of-the-way side halls. Handfuls of shuffling loyalists were peering at the spacecraft, their expressions unmistakably wistful and nostalgic. The looks on their faces reminded me of nothing so much as the way modern Greeks and Italians view archaeological exhibits on the vanished glories of Athens and Rome.

It wasn't supposed to turn out this way. In the mid-1980s, exultant Soviet space planners had laid out ambitious plans for the next 10 to 20 years in space. These centered on the *Mir-2* space station, with its specialized add-on modules and vast solar-power panels, and the *Buran* shuttle, which would bring equipment and supplies to the station. An eight-man transport ship called *Zarya* was being built, based on a scaled-up *Soyuz* design, and it would be carried into space on *Zenit* rockets. Fleets of robotic interplanetary probes would open the way for human flight to the Moon and Mars by the end of the century.

Such expectations, as well as the fears of Western observers, were based on remarkably stubborn illusions about the "momentum" and "continuity" of the Soviet space program. Observers would often project the future as some sort of linear extrapolation of past events. But to use the analogy of a boiling pot of water, when the gas is turned off, the boiling stops with shocking suddenness. In this case, the water stayed hot for a while, but its energetic motion ceased. Analogies can be just as deceptive as self-delusions, but what was so shocking about the collapse of the Soviet space program was how few people noticed it, either inside the USSR or elsewhere.

Like all dramatic historical declines, the slippage of Soviet space ambitions had deep cultural roots and was the result of many different factors. In the late 1980s, even as the Soviet space effort peaked and its greatest technological triumphs occurred, the government had already turned off the gas, in large part because of the widening public disenchantment with useless show-off space shots. Cutting wasteful government spending was one of Prime Minister Mikhail Gorbachev's key strategies. In a speech to the Assembly of Peoples Deputies on May 30, 1989, he called for even more cutbacks: "Expenditure on the space program has already been partially cut," he told the government officials. "We must seek more possibilities in this direction."

Speaking to the Congress of Deputies a few months later, Gorbachev stressed his acceptance of the profit motive for space spending: "The latest developments made during the *Buran* project alone could have significant benefit—worth billions of rubles—if they are passed on to national economic enterprises and organizations," he said. "One must bear in mind that only if that happens will the money we spend to master space be justified."

One disagreeable fact that was hobbling pro-space Soviet officials seeking more government funding was that, aside from the fading propaganda glories, their country had nothing much to show for decades of human space flight. NASA, on the other hand, could argue that the United States had measurably benefited from its space spending. The most obvious benefits were the spectacular scientific breakthroughs, such as the hundreds of kilograms of Moon rocks that revolutionized planetary science and the far-ranging interplanetary robots. Less obvious but much more productive was a technology utilization program and the general stimulation of U.S. industrial capability by the commercial application of techniques learned during work on NASA contracts. True, many famous "spin-offs" are more "spin" than "payoff," but the measurable contributions were genuine.

Not so in the USSR. Its space industry was always so compartmentalized behind walls of military secrecy that little if anything leaked into the commercial sector. Often, potentially useful developments, such as an automated landing system for *Buran*, had to be independently and expensively reinvented by other avionics groups that were cut off from the secret technologies.

By the late 1980s, a number of insightful and courageous Russian space experts were able to speak out in public about this situation. They were interested in finding the practical benefits of space activities.

In 1988, Dr. Vsevolod Avduyevskiy, a leading official at the Institute for Cosmic Research, expressed his frustration with all of the empty talk about "space factories" for unique materials. "Operations involving the industrial production of materials have not yet been started," he wrote, complaining further that "this represents a definite loss of speed." His advice was to make a better connection between research and industry: "Interested agencies should be consulted [to] identify materials that are most promising from the standpoint of industry, and then efforts should be directed toward producing them in orbit in the needed quantities."

Lieutenant General Kerim Kerimov, then 71 and newly retired, had been the officer responsible for the conduct of all Soviet human space missions. When asked about the practical benefits of the missions, he was blunt: "Still very few, although there could be more." As the director of all cosmonaut missions, he was concerned over lost opportunities: "The quite unique research involving materials science, biotechnology, medicinal preparations, and remote sensing of natural resources has yet to find practical applications," he admitted in an interview. "A long-range, goal-oriented program of research to be conducted aboard space complexes has not yet been developed [although] requests come in sporadically."

"We do not always know whether the things our scientists and cosmonauts are working on will have applications in industry," Kerimov complained. His chief of cosmonauts, Vladimir Shatalov, agreed: "Crews bring findings back to Earth, and everything will then disappear 'down a hole' as it were," he told a news reporter in January 1989.

Shatalov, a cosmonaut who had been director of the Gagarin Spaceflight Training Center in Star City since 1971, was equally blunt. "I must say that readers are right who ask this question: why is there such insignificant use made of space flights to resolve urgent national economic tasks? Speaking of human flights, for example, the return on them for our economy could, in my view, be immeasurably greater."

Shatalov mentioned two decades worth of experiments on materials processing under conditions of weightlessness. "Experiments have been completed, but what now?" he asked. "On a permanently manned orbiting station it would be possible to organize semi-industrial production of medicines, crystals, semiconductors, and many other things. But for some reason no one is interested in it. Perhaps we have not yet awakened after the 'stagnation hibernation'?" ("Stagnation" was the code word used to criticize the Soviet economic slowdown under Leonid Brezhnev, who died in 1984.)

When asked about reports of industrial research on American space missions, Shatalov was blunt. "Unfortunately, such tendencies are very weak in our country," he admitted. "We have a big short-fall here. And readers are quite right to note this substantial shortcoming. It is clear that an abnormal situation has taken shape and must be changed as quickly as possible.

"There is a lack of purposefulness and consistency," he continued. "Clearly this is due in some degree to the general state of our country's science, which is poorly coordinated with the economy and is remote from practical needs." He concluded, "It is objectively time to embark on extensive practical work in space. The scientific quest stages have dragged on for too long."

Journalists picked up the theme. "For decades, cosmonautics was considered almost a sacred cow of the economy," wrote *Izvestiya*'s space correspondent, Sergey Leskov, in March 1989. "To argue about the cost of supporting it was considered wrong and unpatriotic."

But with Gorbachev's reforms, just such arguments were turning up everywhere. For example, the vigorously pro-Gorbachev weekly *Arguments and Facts* published a letter from a group of health workers in Shakhty, a mining city in the Ukraine. "When will all this 'peaceful exploration of space' come to an end?" they demanded. "It is impossible to remain undisturbed seeing people's money go down the drain! Come down from the heavens to our sinful earth! There is no sugar, soap, drugs. . . . Shame! Mother Russia has remained as backwards as it used to be!"

On Cosmonautics Day (April 12), the anniversary of Gagarin's first space flight, the labor daily *TRUD* reported that its mail was 10 to 1 in favor of sharp cutbacks in space expenditures. Some readers even called for a complete cutoff of funds, asking, "Why do we need a space program if we don't have sufficient means for priority needs here on Earth, and if millions of people living below the poverty line cannot make ends meet?"

Those letters echo complaints that were heard widely during the national election campaigns in March 1989. Maverick Moscow politician Boris Yeltsin was the loudest, but not the most extreme: "I'm not saying we should abandon space research, just stretch it out," he explained, calling for a five- to seven-year delay in many of the major projects. Yeltsin went on to sweep 90 percent of the vote in the Russian presidential election. Less than three years later, when the USSR disintegrated, he would succeed Gorbachev as national leader.

On the TV program *Moscow News* in August 1989, during a candid roundtable discussion on the value of space exploration, science editor Leonard Nikishin offered a credible explanation for the vehemence of the negative outbursts when he said, "Where do these extreme views come from—the total, angry denial of the need for 'useless' spending on space? I believe that, not least of all, this is a consequence of well-nigh universal irritation over the hullabaloo of many years about our 'space victories.' But the people lived in a different world. They were short of too many good things of life to take these victories close to heart."

True, space exploration had brought substantial military benefits in areas such as reconnaissance, weather forecasting, navigation, and communications. Top Ministry of Defense officials considered their network of space systems to be "force multipliers," enhancing the combat effectiveness of other military branches by a factor of two or three. But decades-old Soviet propaganda spins prevented such values from being presented to the public, who saw "space" as purely scientific in nature and, in the words of Konstantin Feoktistov, a "devourer of immense sums."

The scaled-back plans led to reductions even before the collapse of the Soviet Union, and continued after that event. The early 1990s were filled with Russian press reports about mass layoffs of space workers. Between 1991 and 1994, 115,000 engineering and technical personnel, as well as 90,000 industrial workers, left the aerospace industry. RSC-Energia's payroll dropped from 65,000 to 22,000 employees during the same period. At the *Progress* plant in Samara, where *Soyuz* rockets were assembled, 8000 workers were laid off in December 1994 alone. Yuriy Koptev, head of the Russian Space Agency, estimated in late 1994 that within a year, only 100,000 of the 360,000 workers left in the industry would still have their jobs. An "open letter" appeal from space veterans to Yeltsin said, "Because persons specializing in space activity are not receiving the material support they need, an irreplaceable exodus of highly experienced personnel is occurring."

Dr. Judyth Twigg, an observer of the Soviet and Russian aerospace industry since the early 1980s, testified before Congress in 1998 about the demographic crisis facing Russian space workers. She cited a Russian report that "more than half of the research and design personnel in the aerospace sector are now over the age of 55; about a third are 45–55 years old; and only one percent are under the age of 35." She added that since wages in the space production sector are only three-quarters the national average in all occupations, "even those who remain often spend only a few hours a day at the workplace before turning to second jobs."

All of this remains a daunting challenge today. Someday the old-timers will be replaced, but for the time being, there is a state of fearful uncertainty in the industry over how that can be done. Most of the top space experts are survivors of the generation that worked on *Sputnik* and *Vostok* nearly 40 years ago. Because the same group worked on project after project as the space programs developed over the decades, they are uniquely experienced and knowledgeable. But even though they are all in their sixties and seventies, they rarely document their activities for the reference of future workers. "Vital intergenerational transfers of knowledge about space industry and operations are not systematically taking place," Twigg pointed out.

As a result, an institutional amnesia is creeping in. This is illustrated by a curious incident that occurred at Moscow's mission control center in April 1995. The cosmonauts aboard *Mir* had displayed several pieces of hardware over their television downlink, described the items, and asked what they were for and whether to retain or discard them. After a determined research effort, ground controllers had to tell the cosmonauts that nobody on Earth could recognize the items, much less explain their purpose. There were no written records and no surviving experts to consult. By the mid-1990s, critical archival data had been irretrievably lost. In response to a request from NASA for historical data on failures in one particular subsystem, the Russians admitted that retrieving data for the past 10 years would be "very difficult," and that retrieving older data was simply impossible.

This means that the much-touted mountain of "Russian space experience" was already crumbling, most of it already irretrievable and useless. To make matters worse, more of it was being lost every year. It wasn't just that the past was fading out of reach. The Russian space effort was coming dangerously close to losing its future.

At the dawn of the Space Age, Soviet planners located their future spaceport in an unpopulated region in the heart of their country. For 35 years, the Baykonur Cosmodrome was the springboard of space spectaculars. Poets called it "the shore of the Universe."

The collapse of the USSR in 1991 left the cosmodrome deep inside Kazakhstan, a newly independent nation. Funding for the space center from Moscow quickly dried up. Many civilian workers, unpaid for months, abandoned their posts. Military draftees sent to do the maintenance work rioted and looted, and even the Russian military space agency began withdrawing its space workers because it could not pay them.

By the early 1990s, there was often no heat or running water in the workers' homes, there were no social services such as schools and medical care, and only the drabbest food items were available in the stores. Security collapsed, and Kazakh squatters moved into abandoned city buildings while looters lurked in the city's outskirts. Public health declined rapidly, and diseases spread, especially among the children. The lack of industrial maintenance and trained operators led to a series of deadly disasters, including fires, explosions, and toxic leaks.

Local workers drew up a petition and sent it to the government. "The condition of the people is disastrous," the document stated. "Systems of power, heat, and water supplies, sewers, and telephone communications are worn out. There are neither means nor materials for maintaining them in proper condition. It's cold in the apartments, drinking water is intermittent, and very often, electrical power is off for long periods. The commercial trading network has collapsed. For the inhabitants of the city, life has become a matter of simple self-preservation."

As reported by TASS in December 1993: "A large group of veteran residents of the Kazakh city of Leninsk has in an open letter appealed to

President Nursultan Nazarbayev to impose a temporary moratorium on launching of carrier rockets from the space-vehicle launching site at Baykonur. According to authors of the appeal, this step will expedite search of ways by Russia and Kazakhstan for preserving the space launching site and the city that owes its birth to it. The letter also expresses deep concern over the fate of the Baykonur cosmodrome and the state of its infrastructure."

Journalist Peter de Selding filed a special report from Baykonur for *Space News* in late 1994. "The Russian Military Space Forces stationed here are a dispirited bunch, with widespread complaints that they are no longer given the necessary tools to do the job," he wrote. "Officers complain of breakdowns in supplies of equipment from Russia and other former Soviet republics. They say further that the basic services that make the harsh climate here bearable—adequate schools, heating, electricity, and water supplies—have withered in recent years."

De Selding then quoted an anonymous Russian space manager: "You see all this stuff and you think things in our country are going to crack. And yet we are still here. Work is continuing. Somehow we find a way to keep our space complex working. How long can we keep it up without an improved financial climate? I do not know."

From Almaty, Kazakhstan, came this Western wire service report: "A court in Kazakhstan has sentenced a group of soldiers involved in riots at the former Soviet space center of Baikonur to prison terms of up to 12 years, a news agency reported Thursday. The soldiers were found guilty of setting fire to barracks and looting stores in the town of Leninsk, 2,100 kilometers (1.300 miles) southeast of Moscow, the ITAR-Tass news agency said. About 500 soldiers who were unhappy about military service and had complained of shortages of food and drinking water took part in the riots in June that shook Leninsk, the main town in the sprawling space center."

During several of my own visits to the cosmodrome in the early 1990s, I met and talked with the young soldiers stationed there. They talked freely about their hardships, but their most eloquent testimony was quite unintentional. When asked what small gifts they would appreciate most, they requested not luxury goods such as cigarettes, but far more basic supplies: tinned foods, if any, or pencils (they dared not even hope for pens).

A space launch in October 1994 was almost delayed when thieves stole 3.5 kilometers of communications cable for its scrap copper value. The line was restored on the eve of the launch. A subsequent launch was delayed when thieves tore off metal parts of the *Soyuz* solar panels to resell them as scrap.

The population of Leninsk, once above 100,000, dropped to about 60,000. Reporting from Baykonur, Belgian space expert Theo Pirard wrote for *Satellite News* (September 5, 1994) that approximately 3000 more officers and their families had left over the past two years. All Russian schools in the area were closed because all the teachers had left.

As if all of this weren't enough, there are other problems with the Baykonur site. Ecologically, the region looks to be beyond salvation. As the Aral Sea to its west continues to dry up, dust from the exposed salt flats, laced with decades of pesticide pollution washed down from the cotton fields upstream, is blown by the choking summer winds eastward across the desert to blanket Leninsk. When they were space pioneers, the workers of the city tolerated their own personal hardships, but now, when their children and grandchildren were being poisoned by simple contact with the dirt outdoors, they often were not even able to buy soap to clean them off. Infant mortality and birth defects in the area were unendurably high.

By 1993, I was ready to argue publicly that the cosmodrome was in a state of irreversible collapse (clearly a prediction that was contradicted by subsequent events). This was in opposition to the official U.S. government position, as voiced by Vice President Al Gore on December 15 of that year, that "Baykonur is a world-class facility on a par with any facility in the world."

The health of Baykonur was also defended by Bretton Alexander of the ANSER Corporation, a Washington "think tank," in late 1993. I grew even more skeptical of his conclusions as I identified the numerous factual errors that indicated that he and the other authors really didn't know much about the cosmodrome.

"No trains enter Leninsk," they wrote, but I'd been to the train station and seen differently. They described visiting a launch processing building to see a "*Proton* booster" and its "strap-on boosters," but the *Proton* doesn't have strap-on boosters. It described the controlling authority as the Space Forces, "the fourth branch of the Russian military" (there are five branches of the military structure, but "space forces" are a special subunit reporting to the general staff).

Even the commander of the cosmodrome, General Alexander Shumilin, publicly expressed contempt for the ANSER report. In a Russian newspaper, Shumilin described the visit this way: "After spending less than two days at the cosmodrome and viewing the launch and technical complexes of the *Soyuz*, *Proton*, and *Energiya* carrier rockets (30 minutes at each complex), the guests happily set off home. Not even a high-class specialist is able to assess the technical condition of very complex systems in such a short space of time, apart from forming an impression of their outward appearance. But the U.S. experts drew hasty conclusions without a moment's hesitation: 'the breakup of the USSR has not affected Baykonur's potential.'

"In their haste, the gentlemen from across the ocean did not attach any significance to the following 'trifle': Of the 40 major organizations involved in implementing space programs in past years, just over 10 are left today. The test cosmodrome now functions only thanks to military specialists carrying out their official duty in bleak climatic conditions and in the humiliating position of an indeterminate legal status. The potential of the military contingent, which is fatigued as a result of everyday troubles, is not infinite however. Meanwhile, the U.S. experts maintain that the tragic situation in which the servicemen find themselves has been 'greatly and deliberately' exaggerated. This is allegedly being done for the sole purpose of 'squeezing' more money out of the government."

Shumilin concluded, "We are prepared to meet journalists and experts, specialists and dilettantes. To show and tell them everything the way it is. Our only request is one which Mikhail Bulgakov expressed exhaustively on addressing young litterateurs: 'Write what you see, and do not write what you do not see!'"

ANSER director Dr. John Fabian, a former astronaut, was asked about Shumilin's criticism. He told *Space News* that the report had received good reviews elsewhere in the Russian press, and that "several" positive articles had appeared. After an extensive search, I was unable to find any such article, and neither Dr. Fabian nor Mr. Alexander responded to my written request that they cite one explicitly. It seems that Shumilin's advice on Bulgakov's words was particularly appropriate.

By writing exactly what I did see, I was able to gain respect from Russian space officials who struggled with making the still unseen history more visible to the outside world. Over the years, I developed a productive relationship with the museum at the cosmodrome. Every time I visited, I found that the museum's exhibits were noticeably better. Slava Nechyosa, the director, personally thanked me for publishing stories about the history of Baykonur that had previously been classified and off limits. Each time he got a copy of a new article of mine, he would show it to his commanding officer and request permission to move the relevant artifacts from locked storage into display cases. If it was described in a Western magazine, it couldn't very well still be "top secret," he would argue.

In one display case, Nechyosa had placed several of my books that I'd given him on a previous visit, including *Uncovering Soviet Disasters* and the forged cosmonaut group photographs that I'd found (failed candidates were airbrushed out of later versions). "Even from the next room, I can always tell when visiting Americans reach that case," he told me with a smile. "There's a burst of laughter and a call to bring the cameras." The last I heard, the books were still there, although Nechyosa had retired to somewhere in Russia and I'll probably never see him again.

During one of my inspection tours of Baykonur, I noticed something odd as I walked around my newly renovated hotel in the *Proton* rocket area. On the cement slabs in front of some other residence buildings, I saw signs of recent campfires. This struck me as "cute." I imagined that the temporary residents in these buildings, here for the most recent *Proton* launching a month before, had gathered in the evenings for social pleasures.

The more accurate—and more significant—meaning of these campfires was revealed in a long article in *Rossiyskaya Gazeta*, published during the same week that I was at Baykonur. It described at length the woes of the Russian communications satellite industry, based on interviews with top officials of the Applied Mechanics association in Krasnoyarsk, where Russia's communications satellites are built. The article described in detail a "launch campaign" by institute workers to prepare a communications satellite for launching aboard a *Proton* rocket.

"The famed space center is a pitiful sight today," the article wrote about Baykonur. "Kazakhstan, which charges Russia a huge sum for the lease, is doing nothing to maintain the infrastructure in a more or less decent condition. On the contrary, it has created all the conditions for its ruin. There is no light, water, or heating in apartment blocks, matters have reached the point where association staffers who go there for a launch cook for themselves on a campfire in front of the hotel. They are forced to live in these conditions for three months at a time." I probably saw the remains of those very campfires.

The bottom line was that the Russian space program was enduring great privation and hardship, and that space workers were resorting to desperate measures for survival. But although the engineers suffered, they launched the satellite, and it worked perfectly. Their ability to perform under such conditions astonished and impressed everyone (myself included) who had misjudged the irreversibility of the decline. Their rallying cry, posted on signs throughout the area, was simple: "The difficulties that lie ahead are not as great as those we have already overcome."

4

International Orbits

"On the world stage no country has permanent friends, only permanent interests."
                                        Lord Palmerston

"Men who have worked together to reach the stars are not likely to descend together into the depths of war and desolation."
                                        Lyndon B. Johnson (1958)

Hanging on the wall in my office is a curiously autographed space photograph. Dated 1972, it is the artist's concept of the *Apollo-Soyuz* docking, then scheduled for mid-1975. It is inscribed, "To Jim Oberg from the Apollo-Soyuz Crew," and it is signed by five familiar names. Tom Stafford, Vance Brand, and Deke Slayton would indeed blast off on the *Apollo*. But the two Russian signatures are Vladimir Shatalov and Aleksey Yeliseyev, who had flown in space together three times during the 1969–1971 period.

As part of President Richard Nixon's diplomatic plans for "détente" between the United States and the USSR, serious space cooperation projects were inaugurated in 1972, soon after *Apollo* had won the "Moon race." The idea was to dock two spacecraft with cosmonauts aboard together in orbit to symbolize the lessening tension between the two Cold War adversaries. It would be called the *Apollo-Soyuz* Test Project, or ASTP.

When the mission actually flew, in July 1975, the two cosmonauts on board were Aleksey Leonov and Valeriy Kubasov. Purely by accident, the replacement turned out to be a stroke of luck from a public relations point of view. Instead of two highly competent but colorless cosmonauts with strictly doctrinaire Soviet-era political views, the mission got Leonov, the Hollywood version of what a Russian cosmonaut should look like: competent and courageous, to be sure, but also articulate, artistic, jovial, and philosophic. Kubasov, while less photogenic, also had a subtle wit and charm, as well as cool competence. He later became the only Soviet civilian to ever command a successful space station docking mission.

Leonov, who in 1965 had been the first man to walk in space, had been trained for an entirely different set of missions. In 1968 and 1969, he prepared to command the first Soviet circumlunar mission, which was canceled after *Apollo* got there first. Then he was slated to be the commander of the first three-person crew to inhabit a space station, *Salyut-1*, in June 1971. Days before the launch, however, medical questions about one of his crew (Kubasov, in fact) resulted in the replacement of the entire crew with their backups.

For exactly 23 days, Leonov and his crew lamented their bad luck (the medical issue turned out to have been spurious). Then, on the return trip from space, the crew that had stepped in to replace Leonov's team perished in a freak depressurization accident. Leonov, Kubasov, and a third shipmate had nearly died in their place.

Two further attempts to get Leonov and Kubasov aboard space stations also failed. A rocket blew up during launch in July 1972, and another station tumbled out of control within hours of reaching orbit in May 1973. Only then were Leonov and Kubasov—still cheerful despite all their frustrations—assigned the *Apollo-Soyuz* linkup. It was a consolation prize, but they expected it to be fun. Shatalov, who had originally planned to make the trip himself, realized that his duties as the commander of the cosmonaut training program were more important than another space mission, and it was he who picked Leonov to fly in his place.

All of the difficulties in the actual mission came before launch, which is just the way space engineers like it, since in-flight difficulties can be disastrous. American and Russian space engineers struggled to establish understanding, communications, and trust. Their remarkable success in doing so was demonstrated by the ease with which the actual flight was performed.

The *Soyuz* was launched first, into a low circular orbit. The *Apollo* followed and began a days-long orbital chase that ended when the ships docked together. Using a special intermediate chamber to accommodate different air pressures, astronauts and cosmonauts took turns visiting the others' spacecraft.

The flight was a technological triumph. But its diplomatic significance—the ultimate reason for it to have occurred at all—needs more attention. For NASA, the "space partnership" demonstration was meant to be a pointer toward a happier, safer world, if only everyone back on Earth would follow the examples of the astronauts, cosmonauts, and space engineers from the two countries. Since that didn't happen, these perceptions must have been badly mistaken. It was and still remains necessary to find out what can be learnt from such an experience.

"So how do we judge the success of the joint project?" asked Ed Ezell, author of NASA's official history of the project. "Certainly, we can say that ASTP had a political dimension, one that reflected the improved relationship between the two countries that Presidents Nixon and Ford and Secretary of State Kissinger were seeking." The project had received White House blessing as a symbol of the détente diplomacy of the early 1970s. "As with so many aspects of American national policy, NASA's programs had always reflected the current environment of foreign affairs," Ezell continued. "The next steps in space cooperation would depend upon the international climate."

Ezell described the flight from NASA's point of view: "In April 1976, Tom Stafford noted that the Soviet and American space teams had met all their joint goals—they had designed, developed, and produced the hardware and systems whereby two spacecraft from different traditions could be joined together in space. 'Where both systems were completely separate before,' Stafford said, 'we got together and worked [the differences] out.... the political implications were [such] that we could work in good faith.' Stafford underscored good faith as 'the key to something this technically difficult.' [Program manager] Glynn Lunney agreed with this observation. The real breakthrough made in ASTP was in bringing together teams from the U.S. and U.S.S.R. to 'implement, design, test and finally fly a project of this complexity.' ASTP had been a big job. 'Perhaps we've gotten a bit blase about it...but we [had] an awful lot of hardware that [had] to work well,' Lunney added."

NASA officials also addressed other benefits of the mission." [Johnson Space Center Director Chris] Kraft pointed out that far from being a giveaway project, as many had claimed ASTP to have been, NASA had discovered many things about the Soviet space program that the American agency otherwise probably would not have learned," Ezell explained. "While he conceded that some of this information could have been ferreted out if there had been a reason to do so, both sides had been too busy with their own projects to study in any depth the other's efforts."

But the space partnership, born of and sustained by international relations, also withered and died by international relations. Soviet-supported North Vietnamese troops completed their conquest of South Vietnam only months before the docking. In the months that followed, Soviet-supported Cuban troops were sent to support the pro-Moscow forces in Angola's bloody civil war. Plans were laid in Moscow to assist pro-Soviet forces in Afghanistan to take over the country. These earth-bound events soured thinking about new space cooperation and conclusively demonstrated which end of the stick—diplomacy or space events—was dominant.

Ezell chronicled the changing attitudes: "As detente disappeared from the foreign policy vocabulary, Chris Kraft reflected upon the meaning of these changes for international cooperation in space. 'I guess that you would conjecture that this whole business of the tightening of the belt on both sides relative to each other's exploits in the world of foreign policy these days is certainly bound to rub off on these kinds of negotiations...unfortunate, but a fact of life.' But Kraft was hopeful that ASTP was not the end of cooperation. He thought that the United States and NASA needed to 'continue rubbing elbows with the Russians in a technical space flight sense. And I hope that we can develop a continuing rapport with those people . . . setting goals . . . between ourselves, that we both want to meet, and then working towards them, even if they are long range.'"

Kraft went on: "Now that doesn't mean that we have got to fly in the same spacecraft...together, but if we have a cooperative attitude... and maybe plan some of our work together, I think [it] will lead to a quicker approach to the solution of problems; that would be very beneficial to the world, and certainly has got to be beneficial politically." And George Low, a top official at NASA's Washington headquarters, described another benefit: "We live in a rather dangerous world. Anything that we can do to make it a little less dangerous is worth doing. I think that ASTP was one of those things."

The influence of the 1975 space linkup on international relations remains controversial. Certainly, participants in the project can be forgiven for their enthusiasm about changing the world. But the naïveté of many commentators, who expressed supreme confidence that the space handshake could "teach" other people to overcome history and self-interest and thus become friends, reached the level of self-delusion. It threatened attempts at real diplomacy and jeopardized the development of reliable cooperative projects in the future.

In an analogy from ordinary life, there's confusion about cause and effect when a robin sings in springtime. Is the bird's song a result of the change in the weather, or is it the cause of it? Mistaking the sequence could lead to well-intentioned but stupid birds singing into a snowstorm and freezing to death.

John Logsdon, head of the Space Policy Institute at George Washington University, put this fuzzy thinking into historical perspective in 1988.

"Those that advocate space cooperation as a means of making significant changes in superpower political and military relationships are fighting against most examples provided by history," he wrote. "For most of the twentieth century, a school of international political thought called 'functionalism' has argued for 'peace by pieces'—creating a network of cooperative relationships in specific areas of human activity that would weave a web of interdependence to place constraints on conflicts so they did not erupt into armed hostility.

"Most of the international organization movement in the post-World War I and World War II periods was motivated by this perspective, as have been attempts to foster regional economic and technical cooperation, particularly in Europe. Most students of international politics are skeptical of the 'spillover' argument—that habits of cooperation developed in narrow areas of activity will have impacts in other areas of nation-state relationships."

The "spillover" delusion is illustrated in innumerable quotations from space enthusiasts, politicians (such as the quotation from Lyndon Johnson at the beginning of this chapter), and political commentators. Logsdon (a strong proponent of international space cooperation) warns: "Whether a dramatic, expressive, long-term undertaking like U.S.-Soviet cooperation in human exploration of Mars could transcend functional limits and influence the basic superpower relationship is an interesting speculation, but one that should be assessed with a high degree of skepticism."

The international confrontations of the late 1970s confirmed the skepticism of those observers who were not carried away by enthusiasm for the "space handshake." As new Soviet-fueled wars sprang up in Africa and Asia, even President Jimmy Carter (who had campaigned on a platform that denounced America's traditional "inordinate fear of communism") announced that his eyes had been opened and that the U.S. policy of speaking sweetly to Moscow had not, in fact, made the Soviets act in a friendly way.

Even at the time it occurred, ASTP was becoming a diplomatic anachronism, delivering a message that was already obsolete. Twenty-five years later, observers began to wonder if the same fate was overtaking the Russian role in the *International Space Station*—but that's getting ahead of our story.

Even though it failed to deliver the promised diplomatic benefits, *Apollo-Soyuz* did make one valuable contribution to the future of international space cooperation. This was a bizarre-looking mating mechanism that 20 years later became the standard for shuttle dockings both at *Mir* and at the *International Space Station*. The smiling cosmonaut-to-cosmonaut handshake of ASTP did in fact lay the foundation for real mechanical spaceship-to-spaceship "handshakes." But the evolution of this device went through many stages, once again mirroring U.S.-Russian relations. When factors on Earth were once again ready, the space hardware would be ready for another demonstration.

Given the inherent problems imposed by the laws of physics, it's no surprise that American and Soviet engineers came up with essentially the same design for docking mechanisms in the mid-1960s. Both countries built systems that worked like this:

The active vehicle—the moving one—extends a long, stingerlike probe equipped with capture latches at its tip. On the target vehicle is a cone-shaped receptacle. When the tip of the probe enters the wide end of the cone, it is naturally guided to the back, where another latch mechanism is waiting. The engagement of these two latches is called "soft docking." The docking probe then retracts, drawing the two vehicles together so that facing rings can be latched together for a "hard dock." The docking mechanisms—the conical "drogue" and the spindly probe—are then removed, and a pressurized transfer tunnel opens up so that the crews can float from one vehicle to the other.

This was the basic design used for NASA's *Apollo* lunar missions and for the *Skylab* space station. It also became standard for Soviet vehicles and, with one exception, has served all *Soyuz*, *Progress*, and science module dockings with Russian space stations to this day.

Both sides had played around with other designs. The first space dockings in 1966 were performed by inserting the entire nose of NASA's two-person *Gemini* spacecraft into a collar on the target satellite. For docking by cosmonauts in lunar orbit, the Soviets designed a barbed probe that could penetrate a flat mesh structure and grab hold anywhere on its surface. But both of these special-purpose designs were dead ends.

The inescapably "male" and "female" nature of the probe/drogue system led to countless earthy jests by astronauts and cosmonauts over the years. The major drawback was equally obvious: Only mechanisms of different types could successfully mate. For short space flights this wasn't really a problem, since each vehicle could easily be outfitted with mission-specific hardware. But engineers knew that at some point in the future, spacecraft would need to be able to dock with any other vehicle in orbit.

The "androgynous" docking mechanism sprang from this anticipated requirement. When the Nixon détente thawed relations between Moscow and Washington in 1971–1972, the resulting plan for the symbolic *Apollo-Soyuz* orbital docking gave space engineers the opportunity to build and test an androgynous docking mechanism. The new design had an immediate political advantage: Neither the Soviet nor the American spacecraft would appear to be dominant.

The initial proposal was for a mechanism that would allow a rescue vehicle to dock with either a stranded transfer vehicle or the crew of an isolated space station. The plan was soon abandoned, however, since the time required to prepare a rescue mission would almost certainly be longer than an endangered crew could survive. But by then the androgynous design had matured, and it had several advantages over previous designs. However, there were no plans in either country to adapt the design for actual spacecraft until the international project was approved. In the end, arguably for the wrong reasons, space engineers were allowed to do the right thing.

NASA's virtuoso spaceship inventor Caldwell Johnson (a self-made engineer who had co-designed the *Mercury* capsule in the 1950s) drew up the preliminary plans for the system.

Soviet docking expert Vladimir Syromyatnikov suggested the idea of a symmetric ring-to-ring system. Together, American and Soviet engineers came up with a new design. Each vehicle would be provided with a docking ring that had three open "petals" extending out from it. The petals were for alignment only: They fit slot-and-groove-style between the petals of the other vehicle's ring, so that the facing rings could fit together only in the desired fashion. During docking, the ring on the active vehicle (complete symmetry was sacrificed) would be extended outward on shock absorbers and rammed (slowly!) into the passive vehicle's ring. The petals would then interweave like clasped fingers and guide the two rings to a flush contact. Latches on the active vehicle's docking ring would then catch hooks on the target's ring. Finally, after the motion of initial contact damped out, the extended ring would retract to pull the two vehicles into contact. At that point, the heavy latches around both hatches would engage to achieve a hard docking.

The new system worked well three times during the one mission that it flew (*Apollo-Soyuz*), and its advantages over the probe/drogue system were immediately clear. For one thing, the damping mechanism allowed much more massive vehicles to link up. And although it required greater accuracy in alignment from the pilots, that wasn't seen as a problem.

By the time the Soviets were designing the *Mir* space complex in the mid-1980s, they needed exactly this kind of system to allow the *Buran* space shuttle to mate with the station. The shuttles were too massive for the limited probe/drogue design, and the Soviets would now be using different docking combinations: Soyuz to Mir, Soyuz to Buran, and Buran to Mir. The androgynous system was the only one that could satisfy all of these requirements.

The Soviets called their design APAS, for "androgynous peripheral aggregate of docking" ("docking" in Russian is stykovka). They improved on the Apollo-Soyuz design in several significant ways. First and most visibly, the guide petals were turned inward rather than outward, which allowed for a much larger internal tunnel. Structural latches were placed outside the pressurized tunnel, and there was more space for electrical and hydraulic connections. The result was a complicated system of struts, jackscrews, dampers, and actuators perfectly designed for the Buran / Mir dockings. But the system never got the chance to prove itself in actual flight with a Buran "shuttleski."

Nevertheless, in 1993, the Russian system was tested in space. A Soyuz spacecraft had been equipped with an APAS mechanism, with the intention of launching it into orbit to dock with an automated Buran shuttle. The cosmonauts would spend several days aboard the Buran before undocking and allowing the shuttle to land unpiloted. Though the mission had been canceled, the hardware remained.

An APAS mechanism had also been installed aboard the Mir station, at the end of a module called Kristall, which arrived in 1990. It was there to allow the Buran to dock to Mir. After the Buran was cancelled, the modified Soyuz was reassigned to bring a crew to Mir. The only difference was that the Soyuz used the APAS mechanism to dock to the end of the Kristall module for its six-month stay. And in so doing, it tested the mechanism which two years later received an American space shuttle equipped with a Russian-built APAS.

In the late 1980s, American space designers had been developing their own docking mechanisms for the shuttle and the Freedom space station. Their initial plans involved a mechanism on the shuttle that could hook itself to a deployed trapezelike structure that would then slowly haul the shuttle up to the station's hatch.

The only principle guiding the development of this complicated, clumsy system seemed to be that at all costs it should not in any way resemble the Apollo-Soyuz design. As a result, the device was a source of serious anxiety for American engineers, who doubted that it would ever work. By the early 1990s, however, the political winds had changed, and it was no longer seen as unacceptable for NASA to acknowledge Russian space expertise. After a brief review, the system designed for *Buran-Mir* was adopted for shuttle-*Mir* and for the *International Space Station*, with teams in both countries doing the modification work. For the *Mir* dockings, the Russians built only the mechanisms, which were then bolted to the American space shuttles.

Independent of space dockings, sustained international space programs involving cosmonauts began in the late 1970s, sparked by a new round in the space race. Once again, U.S.-Soviet competitiveness drove the two countries to do the right things for the wrong reasons.

On the American side, the European Space Agency (ESA) was invited to build a science module to be carried up inside the shuttle's payload bay. This was *Spacelab*, one of the shuttle program's grandest successes, a triumph of international cooperation. Naturally, in return for their expenses, the Europeans were promised a number of space flights for their own astronauts. Later, dedicated flights of the *Spacelab* module were conducted for ESA, for the German space agency, and for the Japanese space program, each carrying scientists from one of these entities into space.

The announcement of the *Spacelab* program in 1976 prompted the USSR to upstage it with "foreign partners" of its own. For its first group of guests in 1978, the USSR selected representatives from its most loyal East European allies. Vladimir Remek was the first, a Czech pilot with two notable distinctions: He was the first non-American/non-Soviet to fly in space, and he represented a country that a dozen years later would cease to exist. He was followed by a Polish cosmonaut and an East German cosmonaut.

These flight opportunities arose as a result of an operational feature of the Soviet space program. All of the previous Soviet space stations had had only one docking port. But when the *Salyut-6* space station opened in 1977, with docking ports at either end, much more complicated missions were possible. Now a crew could dock at one end and receive robot supply ships (called *Progress* vehicles) at the opposite port. And if the cosmonauts intended to remain in space longer than the three-month lifetime of their *Soyuz* spacecraft, they could replace it. This was done by a special crew, which flew into orbit just as the original ship's warranty was expiring. They docked the new *Soyuz* to one port, and then, after a few days, they returned to Earth in the old *Soyuz* from the other port. This pattern of "swapping *Soyuzes*" continues to this day, although *Soyuz*'s lifetime has grown from three to seven months.

A single experienced pilot was capable of flying the *Soyuz* on such a simple mission, and this left the second seat in the ship available for somebody else. Although some space officials wanted these missions to carry scientists into space to do research, the Soviet government opted for the more practical alternative: Carry up representatives of "friendly" nations. After 1981, when the *Soyuz* became a three-seater, two guest slots per mission became available.

The guests didn't need much training, either. They needed to know how to handle the basic spacecraft equipment and how to perform a few days' worth of experiments and ceremonial activities. The purely token nature of the program was shown in 1979–1981, when the next groups of guests were flown in precise alphabetical order (according to the Cyrillic alphabet, where V is the third letter): Bulgaria, Vengriya (Hungary), Vietnam, Kuba, Mongolia, and Romania.

The program caught on and was expanded to the nations that Moscow was wooing diplomatically, such as France and India, and later Syria and Afghanistan. By 1990–1991, as budgetary constraints began to strangle the Russian space industry, space officials realized that they could charge money for such seats. There followed a sequence of commercial flights: The seats were occupied by representatives of a Japanese TV network and a British self-financed project, and then by teams of government-funded researchers from Austria, Germany, and France. John Denver had asked the Soviets to take him on a flight, but as one Russian told me, "He wanted to pay for it with a song."

In 1999, the last foreign guest in the series was from Slovakia. This was a new country, formerly the eastern half of Czechoslovakia, which had broken off. People there had reportedly resented the fact that the first Czechoslovakian cosmonaut had come from the *other* half of the country. They had to start their own separate nation to get equality in space.

By the mid-1990s, selling space seats to foreigners had become a major source of revenue. Indonesia, Malaysia, Greece, Finland, and Argentina were approached, but they all declined. South Africa got really excited about the "formal invitation" that it received in 1995, but several months later, when the government realized that it would be billed, it quickly lost interest. As in the Japanese and British deals, Russia did not require that the deal be directly with a government. In 1997, there were discussions of a paying seat for news reporter John Holliman at CNN.

On the American side, international partners came from the *Spacelab* program, from some of the key contributors to space shuttle hardware (such as Canada, which contributed the superb robot arm), and from similar seats-for-services swaps. The nations involved were Germany, Canada (including a man born in Iceland), the Netherlands, Belgium, Switzerland, Italy, and Spain, among others.

Sometimes the foreign astronauts were representatives of the commercial interests that sent communications satellites into orbit on shuttle missions in 1984–1986. When the *Arabsat* was launched, the company selected one of their officials, Sultan al-Saud, to go along (he was in fact a minor Saudi prince, but there are more than 100 of them in various government posts). When *Morelos* was launched for Mexico, Rudolfo Neri went along. The same deal applied to American customers. Both RCA and Hughes had their own representatives, although Greg Jarvis from the American satellite manufacturer Hughes Aerospace got bumped from his original flight by two space-riding members of Congress and wound up aboard the *Challenger* when it exploded. After that explosion, scheduled flights with representatives from India, Great Britain, and Indonesia, and plans for flights of the Secretary of the Air Force and high-ranking generals, were canceled.

Occasionally, the foreign guests were invited for purely political reasons. During a U.S.-French summit meeting in 1983, President Reagan was briefed to request permission from French President François Mitterand to use the Pacific Island of Hao (in French Polynesia) as an emergency landing site for space shuttles launching south from California. Mitterand readily agreed, and then he added that it would be wonderful if a French astronaut could ride on a space shuttle. Reagan approved the deal on the spot, and only later told NASA.

A few months later, two French astronauts arrived at the front gate of NASA's Johnson Space Center in Houston. They had just finished a mission aboard the Russian space station *Salyut-7*, and they were ready for a new challenge. NASA, however, wasn't ready for them. They were left to wait at the gate for hours while officials tried to determine who was supposed to welcome them and where they would be allowed to visit. It was the wrong way to treat anyone, particularly the French. The Germans weren't much happier with their treatment by NASA, privately citing an overwhelming "lack of respect" in the way they were dealt with.

Later came still more political selections for symbolic space flights, astronauts from Ukraine and Israel in particular. Astronauts from Sweden, Brazil, and other space station partners were also selected.

And it hasn't been necessary to cross borders to have an international flavor. Both the American and Soviet domestic space teams included people with foreign connections. U.S. astronauts have been foreign-born citizens (from Rome, Hong Kong, Shanghai, and the U.S. territories Truk and the Panama Canal Zone) and naturalized citizens who were originally from Australia, China, Holland, Costa Rica, Great Britain, French Indochina (he left before it became Vietnam), Spain, and Peru.

Soviet cosmonauts have been natives of several now-independent republics. Foremost, of course, was Russia, including several autonomous non-Slavic regions (one was a Chuvash, another a Lak from Daghestan in the Caucasus). Others were from Ukraine, Belarus, Kazakhstan, Uzbekistan, Latvia (ethnic Russians), and Azerbaidjan. But they all flew under the Soviet flag. Salizhan Sharipov, who flew to *Mir* aboard a space shuttle in 1998, is a citizen of the Russian Federation and a native of independent Kirghizia, but he is an ethnic Uzbek.

It was easy for both the United States and the Soviet Union to accept foreign partners in subordinate roles in their space programs. What proved difficult, after *Apollo-Soyuz*, was finding the appropriate international relationships that would permit a renewed and equal partnership between the two main space powers.

In the early 1980s, with the space shuttle and *Spacelab* flying, American space officials again spoke favorably about building on the precedent of the *Apollo-Soyuz* linkup. Ten years after ASTP, NASA administrator James Beggs was receptive to a shuttle-*Salyut* space docking. "If they were to accept, we could probably work it into our shuttle schedule relatively rapidly," he told a journalist. "We would like to do it, and it would demonstrate our capabilities to work peacefully together in space for humanitarian purposes."

But the diplomatic situation was still unripe. In April 1985, President Reagan had been about to invite Premier Gorbachev to arrange a shuttle-*Salyut* docking, but the offer was angrily removed from his speech when Soviet soldiers shot an American army officer in East Germany and left him to bleed to death on the street. And by then, Reagan's proposals for a national missile defense had outraged Moscow. Foreign Minister Andrey Gromyko laid out the preconditions for partnership in space: "It has been emphasized that USSR-USA cooperation in peaceful exploration of outer space is possible if there is no parallel program of militarization of outer space. One cannot imagine such a situation when the Star Wars program and joint peaceful experiments could be carried out at the same time. Cooperation in such conditions would be a screen, a coverup for plans of militarization of near-Earth space." It should be remembered, as mentioned in Chapter 2, that the USSR was simultaneously making final preparations for a test flight of its *Polyus* space battle station.

On a practical level, some cosmonauts expressed skepticism about the technical feasibility of such joint missions. Speaking at the Boulder Center for Science and Policy, Boulder, Colorado, on April 6, 1987, ace Russian space docking cosmonaut Vladimir Dzhanibekov said, "It is possible, though not probable with respect to the docking procedure. Docking involves precise and complex maneuvers between spacecraft. The U.S. shuttle is not equipped to do these things." But I'd been on duty at Houston Mission Control in 1983–1984 when the shuttle had performed precisely such maneuvers.

But by the late 1980s, after a decade of fruitless speculation wrecked by political realities, the political situation did change. Once again, politics proved to be the driver of space cooperation, not vice versa.

For a conference on human flight to Mars, hosted by the University of Colorado in June 1990, I wrote a paper on what I saw as the realistic prospects for closer U.S.-USSR space relations. Titled "A Near-Term Incremental Strategy for US/USSR Manned Spaceflight Cooperation," the paper was later submitted to NASA's "Augustine Commission," an independent commission that had been asked to consider a variety of space policy options.

I wrote that if opportunity can be measured by the scope of selectable options, then there had never been an era in space exploration more pregnant with possibilities than the current one. This was because "the central axis of world diplomacy, the decades-long US/USSR face-off, was metamorphizing with breathtaking speed."

Not long before, Soviet space science chief Roald Sagdeev and Dr. Carl Sagan had proposed a future joint cosmonaut mission to Mars. Sagan had long been an opponent of human space flight, but he was also interested in easing tensions with the USSR, a goal that apparently trumped his objections to astronauts. Also, the U.S. National Academy of Sciences had released a report calling for smaller initial steps that might eventually lead in that direction.

I catalogued a number of the steps necessary for joint cosmonaut space flight. They included the interoperability of radio communications and space suits, exchanges of orbital data and solar weather observations, and common standards for space medical data. I specially suggested exchanging flight hardware related to the crew systems (such as life support, communications, displays, and control devices) aboard space vehicles.

Such steps would answer some specific questions. We had to determine what the Soviets really had that would be of practical value to our space program. We also needed to find out who was emerging as the new leader of the USSR's fragmented space-flight infrastructure.

Next would come the exchange of astronauts and cosmonauts for "guest flights" on each other's vehicles. I suggested that doctors, mission specialists, and perhaps materials scientists and Earth resources specialists should be the ones to go.

Preliminary space docking tests might involve a *Soyuz* docking to a U.S. shuttle-*Spacelab* mission. These could perform various demonstrations of technology, including tethered operations, in which two vehicles are connected by a strong line hundreds of meters long, or even longer. These would be followed by shuttle dockings to *Mir*.

I speculated that the shuttle could bring up U.S.-equipped modules (the Soviets would provide the basic framework) to be attached to *Mir*. In return, NASA specialists would remain on board *Mir* as part of the long-duration crew. They would then be able to perform experiments that otherwise would have been delayed for years, until the completion of the U.S. station.

To support this, I continued, each country would have to have a full-time office in the other's astronaut flight space center. The issue of preserving national technology secrets would be handled not by trusting in walls but by constantly developing better technology. Experience has shown that the most effective way to maintain a technological edge is not through vain attempts to keep the other guy down but through the much more Yankee urge to stay several orbital leaps ahead of the other guy. Cooperation would help, I predicted: Imaginative ideas would fuel the ingenuity gap, and the hybrid vigor of U.S.-Soviet cross-fertilization promised to engender robust imagination.

Enhanced mutual knowledge would lead to gains for each country. Even better, both sides could lose something by cooperating. They could lose their ignorance and misconceptions about each other, and it would be about time. Ten years later, I still think it's about time for this to happen.

I advised against committing to more serious long-range projects until the political situation in the Soviet Union grew more stable. After all, the staying power of the Gorbachev regime, or of like-minded successors, was unpredictable. I discounted the widespread and foolish advice to rush in and sign advantageous agreements immediately, before the political climate changed. It seemed unlikely that any future regime that wanted to repudiate Gorbachev's policies would be willing to abide by such "scraps of paper."

I therefore recommended parallel, coordinated space activities, rather than a complete melding of the programs. Joint operations would provide mutual assistance and support, and they would allow for a significant relaxation of the stringent reliability levels needed when a space vehicle is entirely on its own. This is the Antarctic model of international cooperation, which consists of independent but mutually supportive bases and transportation infrastructures.

By following this plan, the diplomatic dangers of fully integrated programs would be avoided. If either side were capable of harming the other side's program, the temptation to blackmail would be powerful. I felt that a fully integrated program would be unstable, in the sense that tension would beget more tension, possibly mounting to the breaking point in a very short time. Each party would be tempted to punish the other by withholding crucial segments of the integrated project, or threatening to do so unless it was satisfied with concessions or other resources. In parallel programs, neither side would be able to degrade the other side's programs by withdrawing its support. Such a dual program would be stable; any waverings on one side would tend to create restoring forces that would return that side to the project rather than leave the other to monopolize it. Withdrawing from such a project would harm only the party that pulled out; the party that remained would be unscathed.

Nothing in my proposals struck me as revolutionary, or even particularly insightful. I'd tried to compile the technological opportunities and compare them to the historical precedents. If I forecast any particular hazards, they were only those that Americans had already encountered when dealing with the Soviet Union.

The conflict between the cold-blooded calculus of potential real benefits and the touchy-feely attitude supporting "peace on Earth through friendship in space" has never been resolved. A lot of the debate stems from distorted versions of history. "Remember that time," the commander of Russia's contribution to *Apollo-Soyuz*, Aleksey Leonov, told news reporters in July 2000, referring to the original linkup in 1975, "the insane mistrust, not just for people, but between countries." Leonov bragged that the crew members "discovered kind, good, smart people, who decided to show all of humanity that we are completely different." Leonov was taking justifiable pride in being part of a project that was characterized by friendly international relations, but like so many of the other participants, he saw ASTP outside of its essential diplomatic context, where it was a failure and a detour.

"The idea was to create a symbol of the new thinking," Leonov explained. "It would be a new beginning for both camps."

The thinking exemplified by Leonov's comments totally ignores the historical outcome of the Cold War. When the official Soviet archives were made public, they proved conclusively that the Western mistrust of the Soviet Union wasn't the least bit "insane." It was entirely justified. The "evil empire" of Reagan rhetoric was exactly that, however much the terminology distressed feel-good friendship-mongers. International space ventures based on self-interest and fair exchanges have always been a good idea, but those based on self-delusion and naïve wishes are doomed to frustration and disaster. This was a lesson in international relations that NASA, at least officially, had yet to learn.

5

Origins of
the Partnership

*"One great difference between a wise man and a fool
is, the former only wishes for what he may possibly
obtain, the latter desires impossibilities."*
                                        Democritus

*"The Soviets no longer were a threat in space, and in
the terms that became commonplace among the
veteran ground crews, as well as the astronauts, the
dreamers and builders were replaced by a new wave
of NASA teams, bureaucrats who swayed with the
political winds, sadly short of dreams, drive, and
determination to keep forging outward beyond Earth."*
          Alan Shepard and "Deke" Slayton, *Moonshot*

Inviting the Russians to join in American space operations was an idea
that sprang naturally from the collapse of the Soviet Union in 1991.
As NASA struggled with the out-of-control design for its grandiose
Freedom space station, many experts looked longingly at Russia's
decades-long experience with its own series of small space stations.
Maybe the new leadership in Moscow would be willing to help.

In 1992, President Bush mentioned the possibility of space cooperation with the Russians in written testimony to Congress. House space

subcommittee staffers urged their White House contacts to offer the Russians expanded space cooperation as a reward for political and economic reforms. One agreement that was reached called for exchanging astronauts and cosmonauts in orbit. An American would fly aboard a Russian spacecraft, and a Russian would be launched as part of the crew on an American shuttle.

That same year, Congress directed NASA to evaluate the possibility of using *Mir* as a base for American experiments. The resulting report described an aging space structure that was prone to breakdowns, noise, and vibration; starved for power; and totally inadequate to host any visits by U.S. spacecraft. NASA's experts recommended against the use of *Mir* for any U.S. purposes. But factors other than technology were to be considered in such a decision.

In the early days of the Clinton administration, with political support for the troubled *Freedom* project plummeting, NASA was facing devastating budget cuts. The program had experienced so many cutbacks that space engineers joked that they needed to delete letters from the station's name, that from now on it would be called "Fred." The Russian space program, too, was facing bankruptcy, and Russian plans for a new space station, *Mir-2*, were also threatened. In March 1993, a Russian space official named Yuriy Koptev proposed a solution to this common problem: merging the Russian *Mir-2* program with the American *Freedom* program. He claimed that it could save billions of dollars for both nations.

Lame-duck NASA Administrator Dan Goldin, reportedly fearing replacement at any moment by some "Friend of Bill," responded with enthusiasm. Over the next two weeks, on Goldin's initiative, officials at NASA, the White House, and the departments of State, Defense, and Commerce developed a plan to "invite" Russia to join the space station redesign effort (there was no mention of the fact that it had originally been Russia's idea). Tony Lake, the president's national security adviser, endorsed this suggestion on April 1, 1993. Officials then presented the idea to the president, who was just about to meet Boris Yeltsin for the first time. Three days later, at the Vancouver Summit, Clinton and Yeltsin agreed to the proposal, and officials in both countries were told to "make it happen."

Things had already been happening on the U.S.-Russian space front. U.S.-Russian space business negotiator Jeff Manber recalls how in 1991, he carried "the very first contract between the US government and the Soviets" into Yuriy Semyonov's office at NPO Energia. In the contract, NPO Energia was identified as "a quasi-governmental space organization" of the USSR.

"The contract was historic," Manber wrote, "in that not only was it between an American and a Soviet organization but, equally importantly, it had two objectives: from a programmatic view, it was to study the use of *Soyuz* as an escape vehicle for space station *Freedom*, and hence would lower the cost of space operations for the US. Yet, at the same time, it rewarded the commercialization or privatization of Soviet assets. Very carefully, the officials at NASA had avoided a contractual relationship with an existing government lab or other federal organization, and shared the belief in the importance of Energia as a non-government organization."

"The senior Russian space officials in the room that afternoon were very excited," Manber continued. "To them it meant the start of genuine cooperation. No one wanted a duplication of *Apollo-Soyuz*, a one-shot performed for political purposes. Dr. Semyonov made it very plain in the months to follow that he was willing to embark on a politically bold new path: privatization of his organization in order to reflect both economic realities and the apparent desire of Western governments to engage his industry as market partners."

The United States had unwittingly blundered into the middle of a turf battle inside the Russian space industry. The defense industry council that formerly had run the space program had recently been dissolved, and its function had been assigned to a small new civilian group called the Russian Space Agency (RSA), which was deliberately modeled on NASA to facilitate cooperation with the United States. Yuriy Koptev, a government bureaucrat, was in charge of this group, and he was allied with several government-owned space organizations, particularly the Khrunichev Center (largely derived from the old Chelomey team). The decades-old rivalry between this group and the Korolyov group, now represented by NPO Energia and Yuriy Semyonov, was renewed as the two groups competed for the prestige and profits of a deal with NASA.

Koptev had originally made the proposal to Goldin to combine space stations. He stressed that because his agency was a federal agency, he was the only Russian that Goldin should be talking with directly. At the time, his "agency" had no budget and only a few dozen employees, but its organizational charts looked impressive.

America's top space official promoted the arrangement with Koptev for the joint station, pointing to benefits that transcended science. "There is no event that can better define the coming of the new age than we joining with Russia and actually investing in technology instead of building weapons," Goldin told the New York Times in January of 1994. He painted a grim alternative: "If we don't do this together, then Russia goes its own way and we go our own way." A few months later, at an aerospace forum, Goldin argued that the withdrawal of U.S. support for the Russian role in the International Space Station (ISS) would play into the hands of "radical right-wing Russian space industry" officials who were opposed to Yeltsin's reforms. "We could back away—and we could give the nationalists a self-fulfilling prophecy that will be a disaster to this world—or we can choose to try and support the flicker of democracy in Russia."

Shortly thereafter, Goldin made the same point in more positive terms: "While there are tangible benefits to Russian cooperation, auditors cannot put a price tag on the intangible benefits of international cooperation. It's good foreign policy, and it's good space policy. The Cold War is over, and cooperation with the Russians demonstrates that former adversaries can join forces in a peaceful pursuit which will generate tremendous benefits for both nations."

The president himself became enthusiastic about cooperation with the Russians. On April 20, 1994, Skip Johns, associate director for technology in the White House Office of Science and Technology Policy, spoke at a meeting of the Commercial Space Transportation Advisory Committee. He touted the president's support for the ISS: "I'm looking at a memo of just a couple days ago and he scratched a note on it relative to the station and Russian participation and his comment is, 'Great. It [Russian participation] should help us sell it'."

The formal 700-page U.S.-Russia space treaty was signed in June 1994 at a special White House ceremony, clearly symbolizing the Clinton administration's desire to take credit for it. Said Vice President Al Gore: "After years of competition in space, which symbolized the rivalry between our nations, we have now found a common destiny in cooperation and partnership, a cooperation in space which symbolizes the cooperation we are building here on Earth."

"There are important real benefits for each country," he added later, "in terms of bridges of understanding that develop when we work toward common goals."

Goldin went along with the pretense that the Russian partnership had been Clinton's idea all along. On April 13, 1994, he told Congress: "Let me start by saying that this is a Presidential decision and Presidential policy, and it is viewed to be in the interest of the United States Government to do this in the broader sense."

By that time, and considering the rhetoric, outside observers had a good idea of the actual purposes of the Russian space partnership. The *Wall Street Journal* noted that "Washington's decision to conclude an agreement with the Russians to implement a project to create an orbital station is the basis of an ambitious and risky strategy aimed at consolidating Russia's orientation toward reforms after the U.S. and Western pattern by establishing ties with its military, scientific, and industrial elite." In *Space News*, Andrew Lawler revealed that an inside source told him that NASA was "just a pawn of the State Department," and that American diplomats were more concerned with political benefits than with technical merits.

The diplomats, for their part, didn't want to take credit for the project. "There are very significant foreign policy benefits to Russian participation," noted James Collins, the State Department's senior coordinator in the office of the ambassador-at-large to the former Soviet states. But he insisted that "this administration's decision to proceed with the project was based on its scientific and technical merits."

Not everyone within NASA was as enthusiastic as Goldin and the State Department about the practical value of the Russian partnership. On August 27, 1993, the chief of Mission Operations in Houston, Gene Kranz (the charismatic hero of the *Apollo-13* crisis), sent a memo to Washington describing the significant safety issues "of particular concern" that his team had identified within the joint project. "Agreements established without addressing these issues would be premature," he warned, "and could present problems during future negotiations, or result in a configuration that is complex to assemble and costly to operate." The warning was brushed aside, and within months, Kranz was out of a job—a lesson that was not lost on other officials at NASA.

"The entire concept of *ISS* with the Russians was from the outset a foreign policy decision, with no assessment of the engineering wisdom of what was being done," a senior NASA astronaut explained to me off the record. "To commit NASA's star program to a major change in direction without exhaustive evaluation and open discussion defied logical engineering common sense."

The Russians wanted to be treated as full partners, but they also insisted on being paid as contractors. It was agreed that the Russians would host a series of practice space shuttle dockings to their *Mir* space station, where a few American astronauts would stay for months-long expeditions, and that NASA would pay for this service. These payments were based not on any serious cost-benefit analysis, but on foreign policy considerations. Contemporary events suggested the rationale.

In July 1993, Russia became one of the five states (including China) in the Missile Technology Control Regime (MTCR), which was intended to prevent the spread of missile technology to Third World nations. The Russians agreed to stop exporting cryogenic manufacturing technology to India. Defense, State, and even Commerce Department officials, worried that India might be using hydrogen-fueled rocket engines to build surface-to-surface military missiles, had lobbied for two years against Russia's sale of the enabling technology. But until the space partnership idea emerged, the United States had had no leverage with Russia.

The Russians told the White House that the loss of the Indian deal would cost them several hundred million dollars (other observers considered that number highly inflated). By mid-1993, the directive was clear: The United States had to propose a space agreement with Russia that was worth the same amount.

U.S. diplomats insisted that there was no link between Russia's cancellation of the deal with India and its acceptance of a U.S. space deal with an equal dollar value, but they did admit that "things came together conveniently." Many outside observers assumed that the linkage was direct. The price tag of $400 million for the *Mir* visits alone was otherwise inexplicable. Russian officials valued the Indian deal at about the same amount.

On top of this $400 million was another $200 million to pay for the first Russian-built space station module, the *FGB* (a Russian abbreviation for *Functional Cargo Block*), in return for the right to call it an

"American launch." Miscellaneous hardware purchases and extensions of the *Mir* visits added another $100 million in U.S. payments, bringing the total NASA cash transfer to Russia to $700 million.

"We knew the administration wanted to send money to Russia," a retired NASA official told me in 1997, "but not just as sending dollars. We wanted to get something out of it. But at that time, we were going to send the money anyway." Once the actual figure was set, the next step was to procure enough services from Russia to make the price seem justified.

Ironically, efforts to justify the cost of the program only made it more expensive. A prime instance was the excessive number of shuttle-*Mir* dockings that were to take place before beginning the assembly of the space station. Said Bryan O'Connor, a former astronaut who played a lead-ing role in those 1993 negotiations, "We went and worked out the right numbers, checking with our scientific people and our technical people about the issue of how many times we would like to take the shuttle up to *Mir* to verify hardware, mission control plans, and joint operations." Their conclusion: "We could do everything we needed in four flights."

But while traveling from Washington to Moscow in late 1993, NASA Administrator Goldin told his staff to make it ten flights. "We were completely baffled," recalled O'Connor. "We had to cross out all the numbers on our charts and replace them with the new ones. I didn't know where that idea came from."

The Russians, too, were amazed by the change. "They had worked out the logistics for four flights, and suddenly we told them we wanted to have people on board for two years," O'Connor recalled. "They asked us what was happening to all the science missions that these flights would replace. They asked us why we were trashing our science pro-gram to dock again and again and again with *Mir*." If six more shuttle launches were diverted to docking with *Mir*, their original science pay-loads would have to be canceled. "The Russians thought very highly of the science we were getting from the *Spacelab* flights, they had the highest praise for it. They were just drooling to get on board."

Yet the extra science missions were cancelled in favor of repetitive dockings (at least seven and "up to ten," in the official announcement). If NASA was going to pay the Russians $400 million, the agency wanted it to look like it was getting $400 million worth of dockings, even if it had to cut the shuttle science flight program in half to do it.

NASA's commitment to Koptev's governmental organization "had very concrete results," Jeff Manber recalled, as the old NPO Energia contract was voided and merged into the overall RSA agreement. The results were all bad. "With this single decision," he wrote, "the US became dependent on a space agency that had no government money, no commercial cash flow (hence no ability to fund any future programs), and no realistic control over the programs of interest to NASA. With or without NASA's assistance, RSA was and remains a weak organization.

"During this critical period, senior management of the US space agency failed repeatedly to understand that 'Energia' was not a contractor in the American sense, it was instead the 'commercial' owner and operator of its own programs. Frequent efforts from 1992-94 to relegate Energia to a status equal to that of, say, Rockwell's to the Space Shuttle program [Rockwell built the shuttles under contract to NASA], were offensive to Energia, which had designed, developed, operated, and increasingly funded from their own operations those programs with which it was associated. It created poor relations with the leading Russian organizations based more on a conflict of personalities."

Manber continued: "Given the American response, there seemed to the Russians three prudent paths to carefully explore for the industry: the first [is] the government space agency, which is the Russian Space Agency. The second [is] a quasi-governmental organization, partly government, partly commercial. This is Khrunichev. The third option is RSC Energia, a company with foreign shareholders, private capital, and a commercial mentality. To the surprise of many in Russia, and the disappointment of many in the States, the commercial path was shunned by NASA and the [Clinton] administration, and it was made quite clear that the centralized space agency was the preferred model for the new Russian space program. Ironically, NASA's decision not only not to reward Energia for its commercial efforts, but seemingly to punish the chosen direction, found strong allies in the Russian Duma from the more nationalistic members, who also preferred to keep the industry as close in structure to the communist model [that is, a fully government-controlled entity]. This irony has been lost on NASA but frustrating to those of us seeking to structure the Russian space industry after more commercial Western markets."

NASA's claim that the Russian partnership would make the *ISS* cheaper and faster to build was based on the assumption that the Russians would supply certain essential modules that NASA would otherwise have to build and pay for alone. These included equipment for propulsion and attitude control, equipment for life support inside the station, and a spacecraft capable of evacuating the crew when the space shuttle wasn't docked. In NASA's estimate, the net savings in construction and assembly costs came to $2 billion.

Almost all non-NASA specialists rejected these claims. "I have yet to see a joint international program that saves any money," noted aerospace industry leader Norman Augustine, who had chaired the "Augustine Commission" for NASA a few years earlier. By June 1994, the Government Accounting Office had written: "Most of the savings from Russian participation comes from an optimistic schedule that may not hold up. If the schedule slips, any savings will quickly evaporate." This outside advice proved to be right on target, but at the time, NASA and the White House refused to consider it.

NASA's Goldin responded to the GAO report in a statement issued on June 24: "The fact is every nickel is accounted for in the NASA budget, and Russian cooperation will not cost the U.S. taxpayer one penny more—in fact I believe it will save us billions." Barry Toiv, then a spokesman for the White House's Office of Management and Budget, agreed: "We are confident in our estimate of savings due to Russian participation," Toiv said in the *Houston Chronicle*.

"Russian participation in the Space Station is a good deal for the American taxpayer," Goldin insisted. "It will save hundreds of millions, if not billions of dollars. For the American taxpayer, it's a win-win situation. More space station for less cost. Russian cooperation will not cost the US taxpayer one penny more."

The books were obviously cooked. One crucial gimmick was not counting space shuttle missions as part of the cost of the space station. NASA officials said that this was legitimate, since the shuttle flights, which come from another part of their budget, would have occurred anyway. But in that case, operational costs would have been charged to another program, whose cancellation for the sake of more *ISS* missions was just another hidden cost of the partnership. Omitting the shuttle costs also paved the way for the biggest budget deception of the *ISS* program: hiding the expense of changing the station's orbit.

The original plans had called for the *Freedom* station to be carried up in pieces by shuttles launched due east from the Kennedy Space Center (near Cape Canaveral), taking full advantage of Earth's eastward rotation. The station's orbit would consequently range between 28 degrees North and 28 degrees South latitude (i.e., an orbital inclination to the equator of 28 degrees).

Because of esoteric but immutable laws of celestial mechanics, the Russians, with their far northerly rocket bases, simply could not reach this orbital path. Their missions circled Earth with a much steeper north-south range of 52 degrees. So in order to allow the Russians access to the new space station, NASA shifted its planned orbit northward.

This caused a number of operational difficulties, since NASA engineers had based their designs for the station on the low-inclination orbit. Because the new orbit's angle to the Sun was drastically different, many parts of the station could easily overheat or freeze. Even worse, shuttles heading for the station could no longer fly due east from Florida; instead, they had to head off toward the northeast, losing much of the boost from Earth's eastward spin. Because of this, the shuttle's payload capacity fell by one-third. NASA then implemented a number of design changes to increase the shuttle's payload, but since these changes could have been made no matter which orbit was aimed for, there remained a one-third penalty for the Russian-compatible flight plan.

It's easy to tally up the cost of doing it the Russians' way. Over the planned 20-year life of the *International Space Station*, NASA expects to fly there about 120 times. About 40 of these flights will be needed merely to match the amount of cargo that the first 80 would have been able to carry into the old west-to-east orbit. At an estimated half billion dollars per flight, taking the Russians into the partnership will cost $20 billion. Yet not a penny of this appears in NASA's official space station budget.

The change in orbital inclination had been a feature of the original Russian merger proposal letter back in March 1993, but NASA officials had not drawn attention to it, and Congress was taken by surprise months later. "The controversial thing was not the docking program," recalled Nick Fuhrman, then an aide to the House Subcommittee on Space. "The controversial thing was changing the orbital inclination of the space station." NASA assured Congress that the penalties for the change would be entirely offset by the development of more efficient shuttle launch hardware. While technically true, this claim overlooked the fact that the same developments could have delivered major improvements in the due-eastward launchings as well, improvements that were sacrificed in order to choose an orbit that allowed access by the Russians.

NASA justified the new orbit by pointing out that it allowed the observation of more of Earth's surface, even though the agency had earlier rejected all proposals to do Earth observation research from its space station. The argument was clearly designed with one target in mind: Vice President Gore. "That was a cheap and unfair trick by NASA," recalls Nick Fuhrman, "taking advantage of Gore's well-known environmental inclinations." Gore uncritically accepted this rationale for the awkward northern orbit, but as it turns out, NASA has not funded any significant scientific research that would take advantage of the space station's high-inclination orbit, except for a small instrument for watching sunrises and sunsets. And that unit could just as easily have been hooked to any small automated satellite.

Congress also objected to a station design that allocated critical modules to the Russians, with no backup systems on the U.S. side. Although it was the main basis for the promised cost savings, this dependency worried many members of Congress on both sides of the aisle: Democratic Senator Barbara Mikulski told Goldin that Russia's role should be "enhancing" but not "enabling," and Representative James Sensenbrenner, the Republican chair of the House Science Committee, used critical path terminology to argue that the successful completion of the design should not depend on Russian hardware.

In repeated testimony before Congress, NASA agreed not to put Russia on the critical path, but then the agency proceeded to do just that. In June 1994, President Clinton assured Congress in writing that NASA would "maintain in-line autonomous U.S. life support capability during all stages of Station assembly." NASA did studies of alternative billion-dollar replacement modules, but when it actually tried them in 1997, all of these highly touted contingency plans turned out to be useless. For years, NASA had promised that it had workable alternative plans for space station assembly and operation, in case the Russians reneged on their promises. This was a charade. Seeing that the White

House was committed to keeping the Russians aboard at any price, NASA officials never seriously considered any other possibility.

As envisioned in early 1994, before the signing of the partnership agreement, the Russians' contribution would be extensive. They would build the *ISS*'s first module, the *FGB*, under contract to NASA's station contractor, Boeing. They would finance and build the second module, the *Service Module*, based on their own embryonic *Mir-2* module, and that would carry the station's life support and space-maneuvering systems. They would deliver a string of *Soyuz* space capsules to provide emergency-landing capabilities for the station crew, and they would develop a heavy robot supply ship called the *Progress-M2*, twice the size of existing models, for frequent logistics missions. Follow-on modules would provide more laboratory, power, and operational capabilities. On paper, it was an impressive collection of hardware, and it looked like a bargain—if the Russian promises could be believed.

But in late 1995, the Russians confessed to NASA that many of their promises simply could not be kept. They had no money for the Service Module or any of the follow-on modules, and the proposed heavy supply ship *Progress-M2* turned out to be only a designer's fantasy. Although NASA publicly continued to express confidence in its Russian partners, in private the agency knew better. It began conducting contingency studies to anticipate *Service Module* delays of up to 24 months.

In private briefings to NASA employees, managers passed on the news: "The plan is to let the Russians out of most of their promises," one manager began, according to notes from a listener. The White House had directed NASA not to consider an "all U.S." version with new modules; such a design was "politically unacceptable to the administration." Further, despite the growing evidence of their unreliability, the Russians must be kept on the critical path "to support U.S. diplomatic goals," the NASA official continued.

To preserve the partnership, NASA agreed to shoulder significant new burdens, including two extra shuttle flights to carry up sections of a Russian-built module called the *Science Power Platform*, which Russia couldn't afford to launch on its own. That was an additional billion-dollar expense for the United States. NASA would pay the Russians to redesign their *Soyuz* space capsule so that taller astronauts could fit inside (only half of the American astronauts were short enough to use the capsule). Plans for a joint space suit were canceled, and the old U.S. shuttle suit had to undergo major (and expensive) modifications.

NASA also agreed to pay Russia for two additional *Mir* visits, during which the shuttle would deliver enough supplies to relieve, in Goldin's words, "a significant logistics shortfall." In exchange for this expansion of the original contract and the infusion of much-needed American cash, the Russians made a series of new promises: They would keep on schedule with their own modules, especially the *FGB* and the *Service Module* (April 1998 was the goal). They would also develop a new heavy-class robot supply vehicle to support the new station.

NASA also had to sacrifice some other high-utility plans. It had planned to mount an experimental high-efficiency solar-power module on *Mir* at the end of *Mir*'s crew-related operations, so that the unit could be flight-tested in anticipation of using it on the *ISS*. With the extension crew operations on *Mir*, this opportunity for a useful joint experiment evaporated. But it was no loss, since despite all the costs allegedly saved through the partnership, NASA found that it was running short of money and could no longer afford to develop the new solar dynamic system.

Then the promised new heavy supply ship fell into a black hole and vanished, and Russia told NASA that it could not on its own afford to build enough smaller old-style supply ships.

Fuhrman, the former House Space Subcommittee aide, recalls Congress's frustration upon suddenly discovering that the Russian Space Agency was bankrupt, even though $100 million in NASA funds had flowed into the RSA's New York bank account every year.

"We had expected that the Russians would use the Phase 1 payments to meet Phase 2 performance," he told me. "We believed that this money would go into the aerospace industry that was building their contribution to Phase 2." That should have been more than enough to keep the Russian contributions on schedule. But the Russians told NASA that they were not to link Phase 1 (shuttle-*Mir*) payments with Phase 2 (space station) performance. "We asked Gore staffers what connection there was," Fuhrman says, "and they insisted there was no connection."

Science Committee Chairman Sensenbrenner had tried to find out where the money was going, but the Clinton administration sided with the Russians: "The White House told us not to interfere in the internal workings of foreign governments," Fuhrman said. The administration thus echoed the comments made by a NASA spokeswoman when confronted with the evidence of massive space industry corruption: "What Russia does with their own money is their own business."

As Manber and others had warned, the consequences of NASA's wrong choice of partners and of policies toward the Russian space industry "continued to pile high atop one another during the *Mir* program and the planning for *ISS.*" Manber wrote that "NASA considered the *Mir* program government-to-government. From the Russian perspective it was a commercial contract, signed off by the private corporation Energia. NASA considered the *FGB* module for the *ISS* a component of a government program. From the Russian perspective it too was commercial. NASA considered the service module to be the same, yet from the Russian space industry perspective it could not have been more different: it was truly government-to-government and destined for delays, given the financial collapse of the Russian federation."

The Russians always knew that the *Service Module* was to be their contribution, to be funded entirely from their own resources, but they soon came to realize that not funding it adequately would have no negative consequences. Quite the contrary—it would only motivate NASA to find new ways to pass more money to their space industry.

As they say, you get what you pay for. And you reap what you sow, in space as on Earth.

6

Mir Breakdowns

"Even knowledge has to be in the fashion,
and where it is not, it is wise to affect ignorance."
Baltasar Gracian, *The Art of Worldly Wisdom,* 1647

"If you have a lemon, make lemonade."
Howard Gossage

The men knew that they were in trouble when they realized that their first reactions to the emergency were wrong. When a serious fire broke out on the *Mir* space station on February 23, 1997, the six men aboard—four Russians, a German, and an American—began a struggle for their lives. One of the Russians, familiar with electrical fires from back on Earth, fought back the overwhelming urge to open a window. Jerry Linenger, the American, instinctively sought to get down low on the floor, below the smoke. Within a heartbeat, he realized that there was no "down" in space. The smoke was spreading everywhere evenly.

The crisis caught American space officials off guard—maybe not immediately, but certainly 12 hours or so later when the Russians finally told them about it. And it shocked the world to learn that after so many years of apparently routine space missions, there could still be vicious surprises.

After all, the first three Americans to spend extended periods on *Mir* had had a relatively easy time of it. True, Norm Thagard may have gotten bored waiting for the Russians to deliver his research equipment on a robot freighter in 1995, and Shannon Lucid may have had to spend a few months extra in orbit in 1996 because of a shuttle launch delay, and John Blaha may have had personality clashes with his Russian commander. But they all felt safe in space. As the first Americans to spend more than two weeks in space since the *Skylab* astronauts in 1973–1974, they were reopening a flight regime that would be critical to the success of the *International Space Station*.

But the nation's growing complacency was shattered by the near-catastrophe aboard the Russian station. The crew of six aboard *Mir* faced a ferocious fire that defied conventional wisdom. Its occurrence was completely inconsistent with NASA's image of the Russian space program.

At the root of the fire was the human need to breathe oxygen. In space, oxygen is supplied in various forms, at a rate of about two pounds per day per crew member. Small amounts can be carried in pressurized bottles and bled out through valves and regulators that maintain the proper percentage in the cabin air. Long-term supplies can come from devices that use electric power to break the chemical bond of water—H_2O—and turn it into breathable oxygen and disposable hydrogen. But sometimes there's a need for supplemental oxygen from a source that that is safe and easy to store for long periods of time.

Certain materials give off oxygen during a chemical reaction. They've been used for decades in oxygen masks, in aircraft supplemental breathing systems, and on submarines. On *Mir*, the Russians installed the same system that they regularly used on submarines. Small cartridges of lithium perchlorate in special canisters called solid fuel oxygen generators are activated by striking a pin against a small igniter charge. One cartridge, called a "candle," provides enough oxygen for one crew member for about a day.

At the time of the fire, *Mir* was in the middle of a crew changeover, and three new cosmonauts had just arrived in a *Soyuz*. While overlapping with the old crew for about a week, they would familiarize themselves with the condition of the station and the layout of the equipment and supplies. One of the newcomers, a German scientist on a quick space visit, would then return to Earth with the old crew. The American astronaut, Jerry Linenger, had been dropped off by a space shuttle flight a few weeks before, and he would stay on board until the next shuttle docking, three months in the future.

Mir was equipped with an aging Elektron oxygen producer, but it was operating at reduced efficiency. With *Soyuz* spaceships at both docking ports, there were no visiting *Progress* robot freighters, which normally contained high-pressure oxygen bottles. So the crew routinely activated a series of oxygen candles, about three per day, during the brief dual crew phase. They'd done it hundreds of times before.

But this time, something went spectacularly wrong. A cosmonaut loaded a candle into the unit, activated it, and turned away. Suddenly smoke and then flames began streaming from the device. Recalled Jerry Linenger, "molten metal and sparks exited from the flame." The passageway to one landing craft was blocked by the meter-long torch, which reminded Linenger of a sizzling flame from the space shuttle's solid rocket boosters. It was too painfully bright to look at directly.

"This was an impressive, life-threatening fire in a closed environment," Linenger wrote in his first report. One particularly nasty surprise was the "rapidity and uniform spreading of the smoke . . . far beyond what I would have expected." Later on, he elaborated: The smoke spread "a magnitude faster than I would expect a fire to spread on a space station. The smoke was immediate; it was dense. It was very surprising how fast the smoke spread throughout the complex."

Although there were some official attempts to downplay the incident ("Small Fire Put Out" was the title of the first NASA press release), Linenger's view was different. "Though a severe fire," he wrote only days after the event, "it was in many ways a best case scenario." The flame had been directed away from, not onto, the station's fragile hull, and the normally cluttered passageway had been cleared out only days before. It would have been very much worse if the fire had pierced the station's cardboard-thin aluminum hull. The air would have rushed out so fast that escape would have been impossible.

A few days later, Linenger smuggled home a letter to his wife with the returning Russian crew. She'd only heard NASA's all-is-well version of the incident. "I didn't realize how serious it was until I got the letter," she told the BBC the following year. It turned out that the extinguishers in some of *Mir*'s modules were still fastened to the walls with bolted restraints designed to withstand launching forces. The bolts should have been removed once the modules had been linked to *Mir*'s main section. But the Russians had been so complacent about the danger of fire that somehow neither the cosmonauts nor the experts at Mission Control who scheduled their daily activities ever got around to it. Fortunately, at the time of the fire, the tool kits for removing the bolts were quickly found, and the extinguishers were made available to the crew members with only a few moments' delay.

The fire damage turned out to be minimal. Four cables on the Vozdukh carbon-dioxide removal unit were damaged. One of them, which controlled a vacuum isolation valve, had to be removed, and this disabled automated operations, so that manual control would be required in the future. Also, the plastic switch cover was damaged, but the switch itself wasn't.

Several of the Russians were injured fighting the fire. Station commander Valeriy Korzun suffered burns on several fingers, as well as burns on his chest as white-hot spatters of molten metal burned through his shirt and into his skin. This was almost immediately confirmed by Moscow flight director Viktor Blagov, who told NASA officials that some of the cosmonauts "have light burns on their hands." Later, I heard from NASA doctors that the burns on two of Korzun's fingers were quite serious, raising questions about whether he would be able to don one of his spacesuit gloves for his upcoming flight back to Earth.

Photographs taken by Linenger after the fire show dark green stains on Korzun's fingers, and similar stains also appear on the hands of a second crew member, Aleksandr Kaleri. In another photo, Korzun bares his chest to show the burn spots, daubed with the green medicinal ointment the Russians call *zelyonka*. I saw these photos, but when I requested copies for publication, NASA refused to release them on the direct order of *Mir* operations manager Frank Culbertson. The official grounds were "medical privacy."

"We are thankful that there were no injuries," Culbertson had announced on the day of the fire. "Nobody was hurt, thank God!" was NASA administrator Dan Goldin's comment. With no evidence to the contrary, especially without the photographs, nobody could prove that there was anything wrong with NASA's story.

I was shocked by the fire, but for different reasons from most of my colleagues. What was appalling was not the accident itself, but how NASA officials could act so surprised by it, and how they did their best to misrepresent it. On national television, Goldin was confronted with a description of the flame slicing across the passageway leading to the aft docking port and the *Soyuz* spacecraft that was docked there. He brushed aside the danger with the comment, "One of the Soyuzes was blocked, but the other one was not, so the lives of the astronauts were not in danger at any point." Neither he nor his interviewer seemed to be able to count high enough to realize there were six men on board *Mir*, and that only three men could land in each *Soyuz*.

Goldin downplayed the hazard. "Even a real emergency situation like the onboard fire," he later told Congress, "proved to be easily manageable by the cosmonauts because they were well trained and equipped for such an eventuality, with a nominal reliable way to return to Earth remaining available at all times."

The accident was a total fluke, space safety experts insisted. Tom Stafford, the retired *Apollo* astronaut who headed a special U.S.-Russian space safety panel, downplayed the accident in the following way: "The oxygen-generating canister is the standard canister that is used for Russian submarines, exactly. They have activated well over 10,000 of these canisters, and we had this one failure." A memo by Stephen Tripodi, a flight controller in Houston, echoed this claim: "The safety history of the candles is excellent," he wrote a month after the fire. "Around 2500 others have been safely used in the history of the *Mir* and *Salyut* programs without incident."

I was shocked to see these claims. These statements, and the view that the fire was a rare freak event, were inconsistent with information I had already been collecting. In late 1994, six months before the first NASA astronaut was sent to *Mir*, I had been reviewing safety documents concerning the Russian modules for the *International Space Station*. As a senior operations engineer in the Flight Design and Dynamics Division, I had a wider familiarity than most with the different technical aspects of space hardware for astronauts. My assigned task was to identify safety issues relating to orbital flight—specifically, to rendezvous, docking, and separation. But I tried to keep the big picture in mind as well.

One document from the Russians was devoted to potential hazards on the *Service Module*, the improved version of *Mir* that Russia was building to provide life support and station control capability. As I skimmed through the pages, I came across a section on fire. In it was the first official list I'd ever seen of all the fire incidents aboard all the previous Russian space stations. It was a short list; actually, it was an empty list. The Russians were claiming that there had never been a single fire aboard any Russian space vehicle.

One minor incident involving an electrical short aboard a *Salyut* station was mentioned, but even there, the document asserts, "no fire occurred." Beyond that, there was a terse descriptor: "None." That seriously conflicted with what I'd heard elsewhere: from U.S. intelligence sources, from interviews with Soviet cosmonauts, and from reports of space-to-Earth radio conversations picked up by amateur radio listeners in Europe. Here was the basis of my dilemma: Unlike most of the other NASA engineers working on the project, I had independent access to information related to many of the fundamental assertions that the Russians were making. And my sources were saying things that conflicted with the official Russian version that NASA had instructed us to trust.

Rumors of fires aboard Soviet space stations have been around for a long time. In the early 1970s, I heard a story about a difficult-to-extinguish fire aboard the world's first space station, *Salyut-1*, in 1971. I mentioned the incident in my 1981 book, *Red Star in Orbit*. But it took another decade—and the collapse of communism—before a memoir could be published in a Moscow newspaper about the smoky, smoldering electrical fire that the *Salyut-1* cosmonauts spent hours trying to locate and then control.

Cosmonaut Valeriy Kubasov had been on the prime crew for that mission, but he was dropped at the last moment because of a suspicious spot on his lung. He followed the whole flight from the control center, and in 1992, he further confirmed the occurrence of the fire to Dutch space historian Bert Vis. "It's true there was a small fire on the space station," he explained. "One of the electronic devices started to smoke and they disconnected it. They had fire extinguishers which they sprayed it with."

More details came to light in Asif Siddiqi's monumental book *Challenge to Apollo*. No sooner had *Salyut-1*'s crew boarded the station and turned on the air purification system than a powerful smell of something burning drove them back into their *Soyuz* ferry craft. The next day, the smell was gone and the crew returned to the station. Two weeks later, the burning smell returned more powerfully than before, and again the crew fled to their *Soyuz*. Denied permission to return to Earth immediately, they gingerly returned to the station and switched the power circuits on and off in an attempt to identify the cable that was burning. Eventually the smell faded.

There were stories about later occurrences, too. In 1988, in a small meeting room at NASA's Johnson Space Center, French cosmonaut Jean-Loup Chrétien was sharing his experiences aboard a Russian space station with a group of astronauts (and with me). He showed a series of slides taken on his mission, and discussed what he saw as the significance of the equipment and the procedures.

One slide showed a television monitor that looked like a purchase from a discount electronics store, right down to the holder on the back for the power cord. The monitor's case had nothing high-tech or sophisticated about it; indeed, one astronaut asked Chrétien why the Russians had not even used a fireproof housing.

"They don't see any need for such precautions," Chrétien answered matter-of-factly. "After all, they've had several fires aboard their space stations and found they're easy to put out, they're no big deal." The audience's jaws dropped in unison.

In 1978, two cosmonauts, the first crew aboard a new space station, *Salyut-6*, had been aiming to break the American space endurance record, set on *Skylab*. Cosmonaut Georgiy Grechko was the flight engineer, and 10 years later, while visiting NASA as a tourist, he told me about one of the secrets of the flight.

"I was in a seat by the main control panel," he recalled. When his work schedule showed that it was time to activate a new scientific instrument, he turned on the power switch. "I was looking in front of me," he continued, "and when I turned for something, I couldn't see the other end of the station. It was in smoke!"

Leaving the fire extinguisher behind, Grechko tried to locate the source of the smoke. "I dove into the smoke, and understood it was the scientific device," he explained. He doesn't recall seeing flames, but he clearly saw the source of the smoke. "I simply switched it off and went out of the smoke, because I couldn't breathe in it. I switched on the ventilator, and the smoke got less and less, and everything was all right."

He made no secret of the seriousness of the situation or of the alarm he had felt. When sitting on Earth, he explained, the sight of smoke is not alarming. "But when you are in space! You haven't got a chance to jump out with a parachute! And when you can't see one half of your station because of smoke...." He shrugged his shoulders and smiled.

Grechko laughed again and began to compare his experience with that of the next crew to visit the station. They had had another fire, but this time they went "by the book." They discharged the fire extinguisher, which short-circuited the electronics for several feet in all directions. "In order to save one control panel," Grechko grinned, "they destroyed maybe two or three others."

But that fire, which occurred on September 4, 1977, was more serious than Grechko's. The mission commander was Vladimir Kovalyonok, and he provided the following first-hand account: "We had a navigation complex on board," he explained. "There was a testing program, and without our approval, the control center was turning it on and off."

Kovalyonok and his flight engineer, Aleksandr Ivanchenkov, were exercising after listening to a concert beamed from the ground. Suddenly Ivanchenkov noticed a burst of smoke and flames from the control panel. "We started to fight the fire, switching off a number of systems," Kovalyonok continued. They switched off all the fans "in order to stop the air supply to the fire." Kovalyonok then grabbed the foam extinguisher and sprayed the burning unit.

"We took all the necessary precautions—we were ready to abandon the station," he added. "There were plenty of toxic gases in the air. We used gas masks to continue our work."

But these were anecdotes told in private; they were not written down, and they were not communicated to NASA. In the face of the official Russian silence on the question, folklore and fanciful legends spread. In the late 1980s, Russians privately told American colleagues that fire extinguishers had been discharged six times for "smoke incidents." Soon afterward, a Russian space official admitted to several small electrical fires that soon went out by themselves.

An official at NPO Energia, the Russian company that builds and operates human occupied spacecraft, told foreign visitors in 1992 that on one occasion, the air in a space station had had to be changed after a fire left the station inert. In this case, the mythological event could be easily debunked, since it was a garbled combination of two events. The air had actually been changed aboard *Salyut-5* when it was feared that it had been contaminated by the photographic chemicals in the reconnaissance camera. Also, there really had been a visit to an inert station (see Chapter 1), where an on-board fire was only one of the theories for the station's failure. The problem turned out to have been caused by an electrical short, however.

Many of these rumors were hazy and insubstantial. Then, on October 15, 1994, another fire occurred aboard Russia's *Mir* space station, literally at the moment when I was reviewing the original Russian document on *Service Module* hazards.

A small fire broke out inside an oxygen generator canister. As flames and smoke streamed out of the unit, cosmonaut Valeriy Polyakov grabbed a nearby space uniform and covered the fire while turning off the unit's power. There were no injuries and no damage beyond some seared paint and a ruined uniform.

Then came the really frightening part. Back on Earth, the fire was also covered up. It was not mentioned to the news media, and Russia's new space partners heard only vague stories of "a few sparks," if they heard anything at all. Nobody in NASA's astronaut office seemed to have heard anything.

The full story didn't get out until Russian journalist Konstantin Lantratov wrote about the incident in *Space News*. The article was based on an in-flight radio interview with Polyakov and a postflight face-to-face interview with cosmonaut Talgat Musabayev.

Polyakov stressed that the fire aboard *Mir* was quickly detected and extinguished. "You can stop the sensation here," he urged Lantratov. The fire had been "terminated by quick and decisive actions," in his words.

"We obtained certain experience," Polyakov continued. "This is the most valuable because flight, whether we want this or not, is all the same an ordeal." When an *ABC News* team met with Polyakov in December 1997, he elaborated by saying that he had "often" experienced fires involving the oxygen canisters, but that he had never told the ground, only fellow cosmonauts. "It was so routine, it was no big deal," he shrugged. The main lesson he learned was to keep a wet towel at his side whenever using the unit, and to extinguish any fires as they broke out.

Polyakov's view that the original fire in November 1994 had taught the cosmonauts something useful, however, would prove false. The Russians did what they could to make sure that their space partners never learned anything. They simply never told NASA about the 1994 fire or any of the others.

Since I wanted to write about this incident for a space magazine, I asked NASA for an official answer: Had the Russians provided NASA with any information on the actual fire incidents aboard their space-craft? On December 14, 1994, NASA shuttle-*Mir* official Jim Nise replied, "The Russians have provided information on fire incidents for hardware NASA considers relevant to the safety and reliability of joint U.S.-Russian operations [but he would not specify which hardware that was]. After reviewing this information as well as information pro-vided by the Russians about their on-board fire suppression and warn-ing systems, NASA is satisfied with the safety and reliability of Russian hardware."

It was no accident that the Russians kept their experiences with space fires to themselves. They probably realized correctly that Americans would never be nonchalant about space fire. Just mention the words "fire" and "spacecraft" in the same breath, and we react with an instinctive flinch. Remembering how Grissom, White, and Chaffee died during a launch pad fire in January 1967 still makes us shudder. Even after more than three decades, the horror of the *AS-204* (later called *Apollo-1*) launch pad disaster remains seared into memories, both professional and private.

The Russians, too, had plenty of reasons to fear fire. In 1961, at the height of the hide-the-blemishes communist propaganda, cosmonaut Valentin Bondarenko was killed by a fire at the end of a 10-day ground isolation period in an atmosphere of pure (or at least heavily enriched) oxygen. The fire started when he dropped a swab that had been dipped in alcohol onto a hot plate. Although he got out alive, he died soon after-ward in the hospital. The nature of his death was hidden, and years later, the *Apollo-1* astronauts died within minutes as a flash fire swept through a capsule with a pure oxygen atmosphere. Paying once for such

a tragic oversight is bad enough, but Soviet secrecy and NASA complacency made us pay twice (a program manager told me years later that
had he received news of the fatal Soviet fire, he was certain he would
have paid more attention to fire safety issues for *Apollo*). And Jerry
Linenger would very nearly have to pay a third time.

The two fatal fires in the 1960s were as serious as they were because
of the pure oxygen atmosphere in the cabins. In contrast, all Russian
space stations, and all NASA shuttle flights, use a mixed oxygen-nitro-
gen atmosphere closely approximating normal Earth air. Fire hazards
are less, but clearly, they aren't negligible.

In early 1995, I talked with a NASA safety official about the reports
of Russian fires. He told me that his Russian counterparts did indeed
recall some incidents. In one case, when fire broke out behind a panel,
the crew was ordered to abandon the station, but they decided to stay
and fight the fire.

"But the flight crews don't usually tell them of this kind of stuff," he
told me. "They wind up being surprised by comments made to news-
men or in memoirs much later." The Russian safety officials weren't deliberately withholding information, my friend suggested, they really didn't
have the information. "They don't vigorously debrief the crews," he
explained, "and the crews regularly withhold this kind of information."

When I got nowhere through official channels, I wrote a newspaper article about the issue. This was permitted because I had been officially informed that "fires aboard Russian space vehicles" were *not* a part
of my official duties. Hence, there were no restrictions. The article
appeared in *Space News* in April 1995.

The way I saw it at the time, the big deal was not so much the fires in
space but the smoke screens right here on Earth. I figured that if a topic
as spectacular as spacecraft fires could somehow be off limits to NASA's
curiosity, suspicions should be smoldering about what else we weren't
being told.

In *Space News*, I wrote that I was concerned about the future consequences of NASA's ignorance for the health and well-being of tomorrow's spacefarers. American lives would quite possibly become
dependent on full Russian disclosure of all safety issues, I argued, so the
time for incomplete information was long gone. I concluded that only a
full disclosure of all such incidents, and the resulting countermeasures,

would be enough to foster the development of the earned trust that the *International Space Station* project would need as its foundation. Apparently, the many years of Soviet space experience had taught the Russians to be unconcerned about fires aboard spacecraft. But their overfamiliarity with routine fires had taught them the wrong lesson. "The hazard the Russians failed to appreciate wasn't fire per se but a particularly hazardous material," a NASA expert on life support systems explained to me. "A burning solid-fuel oxygen generator is a totally different animal from a normal fire; the fuel (commonly a 'chlorate candle') is basically solid rocket propellant with an excess of oxydizer."

He added a personal note: "I've got a few chlorate candles in the garage I've fooled around with a bit. The flame is like an oxyacetylene torch, particularly if there's something else combustible around, and it isn't inhibited in 0-G because oxygen doesn't have to diffuse to the flame. When I read your piece, the description of the brilliant flame was very familiar; it looks nothing at all like a normal fire.

"However, the Russians weren't the only ones to fail to recognize they were 'playing with fire,' not by a long shot." he continued. "Essentially the same material destroyed ValuJet flight 592 on May 11, 1996, with 110 people killed. Despite all the finger-pointing later, I've never seen any mention before the crash that oxygen generators could be dangerous."

Bringing the technology back full circle to the American space program, he told me something else I'd never known: "Incidentally, the astronauts experimented with a chlorate candle emergency breathing device," he pointed out. "It is very light and produces oxygen for a full 15 minutes, but it was dropped because they couldn't be sure it wouldn't start a fire."

When NASA's Office of Inspector General (OIG) was conducting its own assessment of NASA's safety evaluations, late in 1997, it found other fire-related documents in NASA files that somehow hadn't been circulated. One memo, dated October 15, 1996 (four months before the *Mir* fire involving Linenger), dealt with fire-emergency training on *Mir*. One unnamed NASA astronaut who had already trained for *Mir* expressed concern that other astronauts might have a hard time locating fire extinguishers because their paint scheme blended into the background, especially in smoky air.

"Upon reviewing this debriefing," wrote OIG official David Cushing, "an outside group applying appropriately rigorous safety standards may have questioned the adequacy of fire procedures and drills, raised questions about the availability and suitability of the fire-fighting equipment, recommended the need for more fire drills, and specifically asked for details related to potential fire hazards." None of this happened because the memo was never made available to any of the independent advisory panels that NASA had set up to assess *Mir* hazards. It would not be the last time that NASA withheld critical safety-related information from the "independent panels" it had set up to review its safety standards.

Cushing's conclusion was blunt: "These issues are better raised before, not after a life-threatening event." But they never were. Jerry Linenger later told me that nobody had ever briefed him, before his visit to *Mir*, about the earlier fires on Russian space stations. Only after he got back, he continued, did he begin to hear the stories.

I was astonished to find that shuttle-*Mir* program manager Culbertson later denied knowing anything about the matter. "Nobody ever told me about earlier fires on *Mir*," he told ABC's Sam Donaldson in early 1998. Neither he nor anyone on his staff admitted to having seen my published articles on the subject two years before.

As I saw it, the only way they could have avoided knowing about the fires was to have consciously decided to remain ignorant. Worse, in order not to interfere with the cooperative projects, they had decided to rely on hope instead of sound hazard analysis. It was worse than merely not wanting to know about hazards. They seemed to want *not* to know about them.

When I raised this issue with my management and in wider discussions with the NASA Inspector General's office, a pattern developed: NASA would use and believe only the official Russian version of information about Russian space technology. It chose *not* to seek outside verification, and it chose to ignore any unofficial information that contradicted the Russian documents. In the areas that fell outside my professional duties (which dealt with orbital flight, rendezvous, docking, and separation), I was told that it was "not in your task description" to comment on any inadequacies in the documentation. The problem was, it seemed that it was and would remain in *nobody's* task description.

NASA's spin on the fire was that the space program was lucky that it had happened. "One of the things we found out because of *Mir* is that we did not have the proper fire protection on *ISS*," NASA chief Daniel Goldin told numerous public gatherings in subsequent months. "Now what would have happened if we didn't go up to *Mir*?" he asked, conjuring up an image of a space station crew killed in a fire that was preventable only through safety measures learned on *Mir*.

Just the opposite actually happened. The Russian presence increased rather than decreased the fire hazard on the *ISS*. As early as 1992, solid-fuel oxygen generators (SFOGs) were suggested for use on the *Freedom* space station, recalled Keith Cowing, a space biologist. "I asked a safety officer from [NASA] if such items could be used to augment existing oxygen supplies in a contingency," he wrote. "He replied that they were too dangerous to even consider using inside a spacecraft and that such 'pyrotechnic devices' were prohibited by safety requirements." But after the Russians were given responsibility for the life support on the *ISS*, and even after the 1997 fire, NASA's safety requirements were modified to allow the devices on board.

If the fires on board *Mir* were such "good news," it's curious that NASA made no mention of a second fire-related incident during the next American astronaut's visit. Mike Foale's laptop computer started smoking and spitting sparks, but he didn't say anything about it until the private postflight debriefing. NASA didn't disclose it to the public, either.

One NASA public white paper in 1997 praised "specific design enhancements and modifications of the *Space Station* and other new knowledge based on shuttle-*Mir* experience." At the top of the list was the assertion that "After the fire aboard *Mir*, software for the *Space Station* was modified so that a single command can stop ventilation between modules." NASA administrator Dan Goldin often pointed to this specific item to prove the value of having had U.S. crew members aboard *Mir*.

Further, according to *ISS* engineers involved in building the NASA *Laboratory Module*, news of the *Mir* fire prompted them to add firewalls (partitions in cable runs to allow adequate concentration of fire-suppressing chemicals) along the standoff conduits that carry cables and plumbing along the length of the module. Steven D. Goo, Boeing's chief space station engineer at NASA's Marshall Space Center in Huntsville, Alabama, told reporters for the McGraw-Hill publication *Aerospace Daily* in November 1997 that the *Mir* fire sent his engineers "back to the drawing board" to improve fire-suppression systems.

However, these descriptions of improvements may be garbled, or at least exaggerated. Also, the role the *Mir* experience played in their development is not so clear-cut. First, according to space station engineers, there still was not going to be a single panic button. Although the fans in the U.S. modules are wired so that a smoke alarm or a thrown switch will trigger a shutdown, the fans and air ducts in the Russian modules are not connected in this way and must be shut off manually (this was confirmed four years later, when a false fire alarm struck the *International Space Station* in March 2001 and there was still no "all fans stop" button to push). A single cutoff button had been featured in the design of *Freedom* nearly 10 years earlier. Another such button had been installed on NASA's *Skylab* space station a quarter-century ago, so the idea is not new.

Nor were the firewall changes on the U.S. *Laboratory Module* for the *ISS* added because of the fire on *Mir*. "It was already in the design," *ISS* operations director Kevin Chilton told me in 1998. "We had a good design." For another article I was writing, I had tried to contact Goo about these conflicting assertions, but he never answered my phone messages or faxes.

Fire wasn't the only problem on board *Mir* in 1997. Potentially poisonous coolant was leaking from corroded lines; computers kept crashing; carbon dioxide purifiers and cabin dehumidifiers broke down again and again. The list of breakdowns grew longer almost daily. NASA was urgently asked to explain how it could be sure that it was safe to keep sending Americans to *Mir*. NASA managers clearly *wanted* to continue the flights, both for the experience to be gained and to show the Russians that NASA was a reliable, courageous partner. But the risks they took—and got away with taking—would turn out to provide little, if any, payoff.

They developed various ways to make their preferred decision look logical. On April 15, for example, there was a teleconference between Russian and American program managers. In a summary of the comments, NASA headquarters official Jesco von Puttkamer (no, *he* didn't give me the copy!) wrote: "[Culbertson] wants the Russians to pres-ent/discuss their mission continuation/termination criteria. On Ryumin's hesitation and comment that he sees no reason to terminate the mission, Frank said, he agrees with that but that they have to show to all other people what supports that decision."

Culbertson elaborated on his request in an "informal note" faxed to Ryumin on April 22: "A part of the American approach to managing a spacecraft in flight is to develop a minimum equipment list which identi-fies the set of hardware that must be operational to initiate or continue the mission. Some of our managers expect us to provide such a list. We under-stand that the Russian side does not manage the *Mir* in that fashion."

The note asked for an explicit description of the operating philoso-phy for all critical components on *Mir*, so that "we can keep our man-agement happy without having to address every failure or change that occurs on the station." In conclusion, he explained, "this activity is designed to show that we are proceeding with safety and health as the number one concern."

The use of the verb *show* doesn't have to imply deliberate false-hoods here, but it's clear from other internal documents that officials at NASA still believed that the problems on *Mir* were flukes and there were no reasons to expect future trouble. A memo from head-quarters, dated April 18, 1997, stated: "No new risks have been iden-tified, and no problems are foreseen." In a Public Affairs Office interview with Scott Gahring, the NASA operations lead in Moscow, posted on NASA's *Mir*/shuttle Web site, when he is asked about *Mir*'s health, he replies: "Everything looks good. The systems are gradually being restored to more acceptable performance levels. It looks like we've gone through the darkest part and we're headed toward the light."

On May 26, 1997, after *Atlantis* returned Jerry Linenger from *Mir*, I took part in a PBS-TV *Newshour* with Charlayne Hunter-Gault. We were discussing future *Mir* problems with several NASA officials. Alan Ladwig, the associate administrator for plans, gave the party line: "We feel a degree of confidence that we have overcome these problems. We are very confident we are operating in a safe manner."

I argued that there was a hope-for-the-best attitude at NASA that had led to a failure to perform classic safety assessments. Through a phenomenon that critics call groupthink, NASA officials barricaded themselves behind their "can-do" enthusiasm and were determined to charge ahead. Ladwig counterattacked: "I really have to take exception to his comment that we have some kind of 'feel good' management structure here to talk about safety. This is an insult to the three astronauts that led those three teams. We resent that."

Ladwig was referring to the result of a safety assessment that had been signed off by American and Russian experts (although never released to the public). It had concluded: "The *Mir* complex is ready to support the beginning of the next increment of the U.S. mission with sufficient systems and redundancy to ensure a safe, healthy, and productive work environment."

Michael Foale was already aboard *Mir* at that point. A taped interview with the astronaut was shown. Asked about the safety of *Mir*, he had replied, "I'm not worried about it. The safety is perfectly assured." That's what NASA's best safety analysts had told him. He was about to find out if it was really true.

7

Mir Screw-Ups

"The important thing is to stop lying to yourself. A man who lies to himself, and believes his own lies, becomes unable to recognize the truth, either in himself or in anyone else."

Dostoyevsky

"Every violation of truth is not only a sort of suicide in the liar, but is a stab at the health of human society."

Emerson

The crew only had a second to be afraid as they caught sight of the off-course spacecraft headed straight toward them. "It was full of menace, like a shark," said Aleksandr Lazutkin, later on. "I watched this black body covered in spots sliding past below me." Then the whole space station shuddered. "As soon as it hit, the fear disappeared," he recalled. "We had to succeed, to survive."

On June 25, 1997, a remotely controlled manual docking procedure for robot cargo vehicles was being tested when the approaching spacecraft went out of control and smashed into one of *Mir*'s modules. As the spacecraft was moving at about 3 meters per second, the hull of the module that was hit, the *Spektr*, was breached, and air began streaming out. Just four months after a fire had threatened Jerry Linenger's life, Michael Foale faced sudden space death of a different kind.

Foale later recalled that the crew on board *Mir* found out about the air leak pretty quickly. Air pressure alarms were going off, ears were popping, and everyone could hear a hissing noise. But the answers to some critical questions remained unknown. Where was the hole? And even if they did figure out which hatch to close, which side of the hatch did they want to wind up on?

The station's equipment made too much noise for the cosmonauts to be able to localize the source of the hiss. Logic dictated that the leak was in the station's main module, which lay directly in the path of the approaching cargo vessel. But at the exact moment of impact, Lazutkin had just happened to glance out a porthole. He saw the vehicle stray past the main module, heading for one of the lateral ones.

Had it not been for that chance discovery, Foale told NASA in debriefing, the crew would not have known which hatches to close. They would have had to abandon the station within minutes. As it was, Foale and Lazutkin raced to that module's hatchway, disconnected the wires and the air vents running through it, and blocked it with an emergency pressure cover.

Program manager Frank Culbertson recalled getting "another one of those middle of the night phone calls which seemed to be coming too regularly." When informed of the collision and the air leak, he later told BBC, his next question was obvious: "How in the world could this possibly have happened?" He was completely surprised by the collision; all of his review teams had assured him that it couldn't happen.

Out in space, *Mir* drifted in utter silence. The power was off-line, and the lights, fans, and all of the other mechanisms were still. The crew said that they found the silence "deafening," even "painful." But Foale looked out a porthole at the bright stars and a glowing curtain of Northern Lights.

He told Lazutkin that no matter how bad a day it had been, this was a beautiful moment, and he asked his Russian friend if he didn't agree. Lazutkin didn't. "It's been a terrible day," was all that he could say.

Vasily Tsibliyev, the mission commander, who had been at the controls at the time of the accident, kept repeating that they were going to kill him, that his career was over. Foale tried to cheer him up, too: "Vasily, the fact that Americans are involved in the program means that they can't just push you aside." He replied morosely, "No, Michael, you don't know our system."

Both men were correct. At first, top Russian space officials did lay all the blame on the two cosmonauts. But it soon turned out that the crew had been attempting a maneuver that was almost certainly doomed to failure. The test plan had been inadequately reviewed, and the crew's training was skimpy, unrealistic, and many months old by the time they actually had to carry out the maneuver. A previous test a few months earlier had barely avoided a similar collision. In the end, Tsibliyev was cleared. Not only did he get his full flight pay, but two years later he became commander of the cosmonaut corps. It took a while, but Foale's instinctive optimism was vindicated.

According to an early report, all five of the Russian pilots who had attempted the maneuver on ground simulators had crashed. Russian officials, however, later denied this and claimed that the pilots who followed the correct procedures were successful in manually docking the *Progress*. The key question seemed to be whether the simulators presented high-fidelity images in which the target was difficult to see against the background of the cloud-covered Earth. If they did not, Tsibliyev's excuse about inadequate training would be confirmed—as it later was.

"The ground controllers made a fundamental error by forgetting about safety during preparations for the docking," a senior expert in cosmonaut training, Rostislav Bogdashevskiy, told a journalist. Russian economics minister Yakov Urinson agreed that "human error on the part of crews and mission controllers, and poor recommendations of designers" were all culpable.

RSC-Energia, the group that builds and operates Russian human piloted spacecraft, reluctantly reached the same conclusion. In a memo to NASA, Energia deputy director Valeriy Ryumin wrote: "One can debate the way this experiment was conducted, and one can criticize the underlying ideology and what seems to be unjustified risk." But he defended the acceptance of simulations and reviews as a basis for the decision to attempt the maneuver in space. "Now we understand that apparently this was insufficient, and that additional constraints should have been introduced that would have at least guaranteed safety," he concluded.

Publicly, Ryumin's U.S. counterpart Frank Culbertson was supportive of his colleague. Combining "the strength of the Russian system and methodology with NASA's [strength]" might "result in a capability exceeding that of either individually," he pointed out at a NASA press conference. "Specifically, the introduction of U.S. experiment planning and trajectory modeling personnel into the Russian process can . . . improve the product of the Russian flight control team."

When I interviewed Culbertson for an article in *Spectrum* magazine, he was more direct: "The Russians tend to sneer at NASA's endless review meetings," he said, "but in this case their own reviews were clearly inadequate." Culbertson was sure that, had they followed the NASA procedure, they would never have approved the test in the first place. "And I think they have come to realize that," he added.

NASA officials had also noticed another disturbing element of the Russian approach to space operations, sort of a "delusion of invulnerability" that prevented them from considering the consequences of "unthinkable" hardware failures. In 1994, one Mission Control Center official had described for me a frustrating exchange about joint space simulations with his Russian counterpart. "They asked why they weren't involved in some of the simulated failures, so our side said OK, give us some candidate failures we can script. They replied that they couldn't think of any.

"Our side suggested one system, they responded that it was redundant and couldn't fail. Our side suggested another system, then another, with the Russian side giving the same response—it would be unrealistic to imagine such specific failures because they couldn't happen." This really aggravated those on the American side, who didn't believe for a moment that there had never been any failures aboard *Mir*. But that was the Russian story, and they were sticking with it, at least until after the string of breakdowns in 1997 opened their eyes.

In life, to have any hope of fixing a problem, you must understand its cause. Sometimes accidents are entirely random and uncontrollable, but more often, they are consequences of our actions, or perhaps of our inactions. Even if an accident is entirely random, one can often "load the dice" and reduce the risk in advance by taking proper steps.

Accidents on *Mir*, and in space in general, are no exception. A proper assessment of cause and effect is essential in lowering future risks by determining which strategies can minimize the factors that lead to accidents. The reaction in 1997, however, was a blanket assumption that the *Mir* accidents that so astonished NASA officials were all due to a single, often uncontrollable and unpredictable, factor. Perhaps the station was too old, or perhaps the cosmonauts were incompetent. Such theories would provide NASA (and Russia) with excuses for failing to foresee and forestall the accidents, but they would also be a recipe for new accidents in the future that could have been avoided if a proper analysis of the old accidents had been made.

For example, following the June 1997 collision and the simplistic "pilot error" explanations that became popular, the news media's characterization of the space mission shifted from the "failure-plagued *Mir*" to the "snake-bit crew." In either case, such oversimplifications in assigning blame tend to frustrate efforts to find the actual root causes and then fix them.

The pattern that I saw pointed to a growing number of errors by specialists at the Russian Mission Control Center in Moscow. This notion ran counter to what top officials in both countries wanted to believe.

Culbertson had seemed to be pretty straightforward with me about the causes of the collision. But I later found a copy of a letter he had written to his Russian counterpart, Valeriy Ryumin. It seemed to suggest a less-than-candid public relations plan. "I have received a report that an official announcement will be made this week concerning the incident of the *Progress* colliding with the *Mir*," he wrote. "I would like to make sure that before a final report is released, we have the opportunity to jointly discuss it and to resolve any questions we may have on contributing causes. I strongly recommend against stating causal factors publicly until the reports are endorsed by your highest authority."

NASA spent a lot of effort protecting "proprietary information" about this and other *Mir* problems. For example, one NASA engineer wrote a detailed report on a July 3 teleconference with Russian rendezvous experts, in which the docking plans and the possible causes of the accident were discussed. The message was e-mailed to other specialists at NASA, which was the proper procedure. But it was immediately followed by a message retracting it. "I just received a warning," the engineer announced, that the report contained "information not available to the general public."

With proper information and standard safety analysis, many of the *Mir* accidents of 1997 wouldn't even have happened, or at least would have been far less severe. Culbertson knew this, too.

The following year, he told the BBC, "I still blame myself for not being more inquisitive about the details of what they were doing, and what the safeguards were." He wasn't alone, as I already knew. Others within NASA also blamed him for overenthusiasm, for the classic NASA attitude that "we can do the impossible."

Culbertson had been right about Russia's need for the more formalistic NASA approach to assessing hazards. It turned out that a routine safety assessment that I had done years earlier was the proof of this! But Culbertson was only half right. What he didn't say was that NASA, too, needed to take advantage of the formalistic safety procedures that it had developed long ago, but had set aside for shuttle-*Mir*.

Three years before the *Mir* collision, while performing my duties as the safety assessor for my flight control division, I decided to apply the standard NASA safety procedures to some of the Russian descriptions of their own hazard analysis of the future *Service Module* for the *ISS*. It wasn't exactly "rocket science," but I admit that I cheated a little: I used some independent information that I'd acquired about Russian space experiences to calibrate the official documents.

On October 12, 1994, I'd filed a report on my review of a recent document from RSC-Energia called "Analysis of Off-Nominal Situations Associated with the Service Module." There were 38 contingencies listed, and many of them dealt with my division's concerns about space debris or the rendezvous and docking of space vehicles.

The first thing I noted was that the Russian report provided incomplete and misleading documentation of the actual flights that had taken place in the Soviet program. I suggested that this practice of withholding relevant safety-related flight data was distressing. I also suggested that the Russians should be confronted on this issue.

My first specific comment concerned pressure hull rupture during docking. The Russian description of the failure stated that since the speed of the approaching vehicle is less than 0.3 meter per second (1 foot per second), "the probability of penetrating the SM shell is rather small." But this contingency was incomplete. A sensor/control failure during docking could lead to a faster approach, thereby increasing the danger of hull rupture. I added that past Soviet docking contingencies had resulted in close fly-by near misses with rates in the 10 to 15 feet per second range.

No, I wasn't predicting the *Progress-Mir* collision, which did occur at about 10 feet per second. I was pointing out the flaws in the Russian attitude toward the threat of such an event. I was simply assessing possible hazards using classic space industry methods. Sadly, nobody was able to make any practical use of my memo, since the safety of *Mir* was by decree entirely in Russian hands.

Nor had I issued dozens of different warnings, which would have guaranteed that by random chance some would "come true." My one other concern involved another overlooked hazard. In the memo, I pointed out that the list of failures did not include any discussion of the volume of vehicle-generated objects jettisoned deliberately or accidentally during flight, with the potential for recontact. From my experience, I knew that the damage done by recontact is less a matter of a penetration of the pressure hull than of a jamming of exterior mechanisms. *Salyut-Mir* vehicles generated significant amounts of debris, and on at least one occasion, one such piece (a trash bag) jammed a docking port and required a contingency space walk to clear it. (In March 2001 a spacewalking astronaut dropped a small work platform outside the *International Space Station*, and it drifted off into its own orbit. When it circled back several days later, the astronauts were forced to fire thrusters on the docked space shuttle to move the station out of its way.)

My concern for hazards from nearby free-floating low-relative-velocity objects sparked another personal campaign. It involved the effectiveness of visual monitoring systems on board space stations. The need for a complete view of a station's environment (we'll return to the subject shortly) was underscored by the *Progress* collision when people at NASA began to realize that the collision hadn't been the clean bang-and-bounce scenario being portrayed on national television.

I first heard about the additional impacts on the *Mir* structure in an e-mail from a friend at NASA. "Word is that the *Progress* may have struck *Kvant-1* solar panel #1 (on the Priroda side) while outbound after striking *Spektr*," he wrote; the modules named were various radially mounted laboratories protruding from the *Mir*'s central core module. I sent out a query to several other old friends, and I soon received confirmation. "The Russians have acknowledged additional hits, including the *Kvant-1* array, but do not consider them serious since the array is still delivering electrical power," another worker told me. "You know, like the old NBA ref's saying, 'No blood, no foul.'" In another reply, a friend described how nobody had seemed interested in this possibility when it was first considered: "We pointed out the bouncing from the *Progress* video back in June, but nobody wanted to talk about it," he wrote. "They were so absorbed with the *Spektr* damage. It seemed like everyone was afraid to ask the Russians about it, and they were being very myopic."

A week later, at a routine shuttle-*Mir* press conference, Frank Culbertson was asked whether *Progress* could have hit something else, other than just *Spektr:* "Well, that was pretty clear shortly after the accident when we first saw the video together," he admitted immediately. "I think everybody could see that there potentially were places where it had some type of contact that slowed down the rotation. The Russians also announced shortly afterward that they suspected it might be the case." The video he was referring to, taken from a camera on the approaching supply ship, was not available to the public. According to NASA spokesman Rob Navias, it had been impounded "pending completion of the accident investigation," and I wasn't allowed to see it.

Culbertson also confirmed that nobody could tell what kind of damage might have occurred. "The potential exists that there may be an antenna damaged because the signal strength on one of them is slightly lower than what they had experienced in the past," he continued. "But until they get a chance to actually look at it from outside, we won't know for sure." Until asked, he hadn't seen the need to disclose this to the public.

The danger of a lack of good all-around viewing was brought home six months later, in December 1997, during experiments with a small German-built subsatellite called *Inspektor*. This innovative and potentially revolutionary free-flier was supposed to fly around *Mir* and take photographs, all under the control of the cosmonauts inside *Mir*. If it worked, it would pave the way for more sophisticated free-fliers to aid in the construction and operation of the *International Space Station*.

The 70-kilogram steamer-trunk-sized space robot was mounted in the nose of a *Progress* freighter. After the *Progress* undocked and moved about 500 meters away, the satellite was spring-ejected from its storage canister. It was supposed to begin circling the *Progress*, and then it should have proceeded toward *Mir* itself. But as soon as the satellite sprang free, its navigation sensors failed. It headed off in the wrong direction.

Before the mission, NASA had been told that the Russians had two safety features built into the *Inspektor* program. First, if no further control rocket firings occurred, the satellite would be ejected into an orbit that would naturally drift away from *Mir*. Second, *Mir* could nudge itself out of the way by firing a bank of steering thrusters that would result in a small cross-coupled force. This term refers to leftover thrusting that occurs not precisely in the desired direction. Although it is usually a problem, the effect can in certain cases (like this one) be advantageous.

But the free-flying vehicle did not fail passively. Instead, it fired its thrusters for a few moments, thereby pushing itself onto an unplanned path that could well have led it back toward *Mir*. So the Russians decided to perform their "run away" maneuver. However, according to the daily NASA "*Mir* Status" report (an internal document, not for public release), "there was a formatting problem in software uplinked to the *Mir*, and this prevented the [maneuver] from occurring." Mission Control told the cosmonauts to keep an eye on the *Inspektor* and that they'd think about trying the burn again the following morning.

The following day, December 18, one of the cosmonauts aboard *Mir* was startled by what he saw when he looked outside. "Pasha, a bit excited, points out the window at the *Inspector* satellite which he deployed yesterday," noted the NASA astronaut who was on board at the time. The satellite was clearly not where he had expected to see it. NASA's internal daily report simply stated, "The *Inspector* payload motion relative from *Mir* changed through the night from trailing *Mir* to leading *Mir*." To move from a point behind *Mir* to a point ahead of *Mir*, the satellite obviously had had to pass very close to *Mir*. But nobody knew how close because nobody had been able to see it out of *Mir*'s windows.

The next day, *Mir* performed two separation maneuvers in order to avoid the possibility of a collision. As overheard by radio listeners in Western Europe, the cosmonauts kept measuring the steadily growing range until, at about 800 meters, the *Inspektor* could no longer be seen from any of the station's portholes. NASA's weekly *Mir* report merely stated that the "*Inspector* was allowed to drift away from the *Mir*." There was no mention of the unseen close pass or the failed separation maneuver.

Now fast forward several years. Crew members in flight appeared to share my concerns about viewing capabilities. About a month after the first crew arrived on the *International Space Station*, I heard one of them remark by radio that the *ISS* was just as bad as *Mir*. "We told them we needed more windows on the station, in all directions," the cosmonaut complained on December 2, 2000. "On the ceiling there is not a single window." If they can wait, a special all-window module called a cupola may eventually be attached to *ISS* to relieve this continuing limitation. But as a temporary measure, the *ISS* began to use cameras mounted on the Canadian robot arm that was installed in April 2001 to occasionally obtain spectacular outside views.

In trying to learn from the 1997 accidents so they wouldn't be repeated, it was necessary to understand their root causes. But NASA's continued insistence that the accidents were just random failures, without detectable or preventable causes, made this impossible. On September 25, Thomas Young, a senior NASA engineer who had reviewed the earlier safety reports, told journalists, "Both the fire and the decompression event, they're not aging events. They could have occurred at any time in the mission." Another official asserted that the collision "could have happened on the first day of *Mir's* mission" (in 1986).

This view totally overlooked the circumstances that made the docking test necessary in the first place: The collapse of the Russian space infrastructure was creating demands that were very difficult to meet. A replacement had to be found for the guidance hardware because it was no longer available from the Ukrainian factory that had previously made it. Such an accident would have been inconceivable 10 years earlier, because the robust Soviet space program was devoting a great deal more money and human resources to space at that time than it was devoting in 1997.

And NASA knew better. In the minutes of the July 23–24, 1997, meeting of the Stafford Committee, the following paragraph shows that Culbertson, for one, knew the real story: "[He] said that one reason that RSA wanted to move from the [old] docking system to [the new one] is because they are afraid that spare parts will not be available for the [old] system due to the fact that it is a Ukrainian-supplied system rather than a Russian system." So the reassurances for public consumption were clearly known to be false.

But even if the "blame-the-cosmonauts" game was rigged, that still didn't mean that human error wasn't involved somewhere. And it turned out that the eagerness of the officials in Moscow's Mission Control Center (called TsUP, its abbreviation in Russian) to point fingers at the flight crew may have been a tactic, conscious or not, to deflect attention from a growing pattern of human error within the control center itself.

On July 17, *Mir* again suffered a power failure and went into a slow tumble. This time, the initial story was that one of the cosmonauts— Lazutkin, it would turn out—had pulled a wrong cable during a computer reconfiguration, thus causing the entire breakdown. Again, it was "the crew's fault."

This was the version that was put forth by the Russians, picked up by the news media, and soon endorsed by NASA. It was meant to show that nothing systemic was wrong, that there was nothing to worry about if only those careless cosmonauts could get their act together. The story was a convenient fable from beginning to end.

What actually happened was more complex and much more troubling. A minor error made by a cosmonaut on *Mir* was compounded by a series of misjudgments at Mission Control in Moscow, and a major crisis was the result. The problem wasn't with *Mir* hardware, or even with the competence of the fatigued and confused cosmonauts on board the trouble-ridden station. The problem was with Russia's capability to correctly assess and respond to emergencies in space. That should explain why neither Russian nor American space officials wanted the full story to get out.

Fortunately, there was an honest witness to the drama in Moscow: a flight director from Houston named Philip Engelauf. Over the years that I worked with him at Mission Control, I found him to be possessed of a first-class mind and a calm disposition. Though he sometimes appeared a bit stiff and humorless, his integrity and judgment were unquestionable. Fortuitously, a memo that he sent to his colleagues at Mission Control in Houston fell into my hands.

"The power loss incident starts with what I think was a recoverable error (unplugging a key attitude sensor), but inaction by the ground allows the situation to deteriorate into a major problem," Engelauf wrote in his introduction. "It appears the ground may have aggravated the situation with an incorrect [navigation] uplink." Engelauf reviewed air-to-ground transcripts, interviewed Russian officials, and queried systems experts in Moscow, which led him to a number of "observations" that he wanted to document for his associates in Houston.

As Engelauf and some associates arrived at the TsUP on the morning of July 18, they were greeted by Shift Flight Director Viktor Chadrin. "The crew put us in a corner again last night!" Chadrin complained, although he did admit that the previous shift had not done a very good job of dealing with the incident. Overall, he still showed what Engelauf called "a predisposition to attribute most problems to onboard crew error."

A review of the console logs allowed Engelauf to reconstruct the sequence of events. It was late in the evening, and both the crew and the ground controllers were tired. Russian practice is for flight control teams to work a 24-hour shift, 8 A.M. to 8 A.M., followed by three days off. Although they are expected to rest during their time off, many of the flight controllers have to take second jobs in order to make enough money for their families to get by.

The minute-by-minute summary of events went something like this:

At 8:20 P.M., during a 15-minute pass across Russia, the crew told the TsUP that they were disconnecting some computer cables from an uplinked radiogram, in preparation for a space walk. Later, Lazutkin, who was working on the cables, put it succinctly: "I was not paying attention, I was not looking at the right page."

Mir then circled Earth, out of contact with Russia (the relay satellites which once provided uninterrupted communications had long since broken down and couldn't be replaced). At 9:56 P.M., as soon as Mir was back within range, the crew reported a "situation" involving the computer's switching to an unfamiliar control mode. They admitted that they "must have disconnected the wrong cable."

In Moscow, TsUP operators were unconcerned. They told the crew to calm down until the telemetry data were received. The crew expressed worry, and asked if they should activate their steering jets. Moscow told them to "take it easy" and said that things "look good," although the operators admitted that there were funny indicators in the telemetry data. *Mir* then passed out of range.

At 11:33 P.M., another communication pass began. The cosmonauts reported on the station's attitude, and the ground admitted that it was having difficulty understanding the telemetry data. The next pass over Russia was through a zone where the ground sites had been shut down, so communications wouldn't be restored until 2:42 A.M. TsUP asked that one of the cosmonauts stay awake to talk to them then.

During that pass, the ground expressed more confusion about what was happening. Tsibliyev replied that he really didn't understand what *Mir* was doing, and that the computer wasn't responding to inputs as he expected. Tsibliyev reviewed the cable arrangements that had led to the problem. "The ground is totally surprised and asks what cable they are talking about," Engelauf wrote. "This yields a discussion in which TsUP finally grasps the situation on board."

During the 4:03 A.M. pass, the ground controllers told the crew that they still couldn't figure out the telemetry. They asked the crew to wait until morning for the next shift to show up. TsUP also sent up an attitude maneuver for *Mir* to attempt.

"Several things were surprising in this exchange," Engelauf wrote. "First, although the crew tells the ground immediately that they disconnected the wrong cable and now the vehicle is acting unusually, it never registers with anyone on the ground." It took hours for them to discover the source of the problem, and even then it was "only because the crew tells them." Engelauf was also distressed by the TsUP's lack of concern over the *Mir* commander's concerns.

"Last," he continued, "even though the ground had observed this identical situation as recently as March, the signature was not recognized even after three passes of telemetry. In the March incident an angular rate sensor failed, in this case the sensor cable was disconnected, but the effect was identical. In March, the vehicle was out of comm so no ground intervention was available, and power was lost in that incident as well."

Engelauf roasted the Russian controllers for not identifying a discrepancy between *Mir*'s actual attitude (the way it was pointed in space) and what they had assumed. Once the power crisis took hold, "a graceful shutdown of onboard systems was not fully achieved—the gyrodynes eventually were knocked off line due to lack of power, resulting in physical damage." During the subsequent repowering of the station, "recharging was delayed because . . . still-powered equipment consumed too much of the power, resulting in little or no excess power to the batteries."

In his conclusions, Engelauf wrote that "there appears to be an inability on the part of the TsUP, even when telemetry is available, to identify even major problems, like a loss of a major attitude sensor component or loss of navigation base reference." Further, "the ground does not appear to give credence to an evident state of concern on the part of the mission commander. The sense of team cohesiveness between the ground and onboard crew, to which we are accustomed, is absent." He also criticized the TsUP for a lack of what NASA calls "situational awareness"—a knowledge of the status of all systems.

Engelauf suggested that his team question the ability of the current TsUP team to adequately respond to future emergencies, both on *Mir* and on the *ISS*.

These were the contents of the memo that later leaked. At a regularly scheduled NASA press conference, I asked Engelauf about it. "If you think I was critical in that case," he replied with a wry grin, "you should see me debriefing my *own* teams." I reminded him that I had been on some of those teams, and he grinned again.

The number of times that situations were made worse by the errors of the TsUP in Moscow was really alarming. It wasn't *Mir*'s fault. The TsUP was taking risks and making mistakes at what I considered to be an appalling rate. And that rate had been steadily rising, year after year.

This rising tide of errors came as no surprise to careful observers. Following an inspection visit to Russian space facilities in early 1995, I had written a major report on the visit for *SPECTRUM* magazine, published in the December issue. Concerns about the quality of the TsUP were present even then. *Mir* itself had been feeling the budgetary stranglehold that year, especially with respect to the loss of experienced personnel. I described a case in which overworked and underprepared ground controllers had tried to deploy a solar panel on the new *Spektr* module. Accidentally, they had sent up the unlatching commands in the wrong order. As a result, the panel had jammed, and a risky space walk was required in order to deploy it. But as the space-walking cosmonauts were trying to install a restraining strap on a piece of equipment, they were out of radio contact with Moscow. With no guidance, they had to guess at the proper orientation. Their guess was wrong, and the strap later jammed still another solar panel.

This was in 1995. Problems on *Mir* began to cascade after that. A subsequent space walk to fix the first problem was aborted when it turned out that the water-cooling loops inside the commander's space suit had been improperly configured, again thanks to inadequate ground monitoring. Eventually, everything got fixed. But I speculated on the meaning of the pattern. I had warned that such an increase in errors was yet another sign that thin human resources were being stretched to the limit in a troubled system that was near collapse. I added that not even the Russians knew how much further the system could be stressed before something catastrophic happened.

This wasn't like the *Mir* hardware safety issues, where NASA officials contrived to remain blissfully ignorant about potential hazards. They knew about this issue. In another briefing to the Stafford Task Force in 1997, Culbertson listed "specific instances" (the minutes provided no further details) in which TsUP had made mistakes. "The Task Force agreed that TsUP response to crew situations was not adequate to assure that the best possible actions were always taken in response to systems anomalies, incidents, crises, or emergencies," the minutes of one meeting stated. In public, however, NASA continued to express full confidence in its Russian partners.

And new astronaut visits to *Mir* continued to be approved.

Cosmonaut Vasiliy Tsibliyev, the commander of *Mir* during the collision and other recent troubles, had this to say: "All this happens because of the economic difficulties. Even when you ask them to send something of vital importance for the station, and we are not talking about coffee, tea, or let us say, milk, they cannot do this simply because they do not have it on Earth. Plants are standing idle, or suffering from not receiving sufficient supplies or are asking exorbitant prices. Therefore a lot of things are still missing on the station."

According to Viktor Blagov, Russia's deputy chief flight director in Moscow Mission Control, "All these malfunctions can be traced to the years when the industry has had absolutely no money, and we have had to find ways to survive using spare parts and old techniques." He once reportedly explained *Mir*'s computer breakdowns to a Reuters reporter as follows: "Due to financing problems, we have to use them till they die.... We are saving a lot of money on this scheme, but we really have to decide soon whether we need safety or money-saving."

Hazards to *Mir* came from a mix of equipment degradation and human error, both usually attributable to financing failures that could be measured quantitatively. But Blagov's approach to assessing the level of these hazards was supremely intuitive, as he explained to an American journalist in December 1997. "Our main conclusion is that *Mir* station has not yet reached the stage of, as we call it, the avalanching destruction, as very frequent emergencies are to precede that stage," he explained. "Judging by the experience of *Salyut* operation, the number of emergencies drop drastically after the beginning and remains to some minimal level for a long period of time—and just before the 'avalanche' it increases.

"That is a sign to pay serious attention to the problem. Today we do not see that rise, and that assures us that the station can continue to operate in future. But that is a very delicate issue, you must use your intuition to smell the moment when danger comes and terminate the operation, without crossing the danger threshold.

"All the discussions in mass media that we've passed that threshold long time ago are wrong, as I see it," he concluded. And indeed, *Mir* continued to operate for more than three years after the 1997 crises.

Intuition aside, the way Blagov wanted to remember history is not the way it really happened. The earlier *Salyuts* were shut down when they ran out of rocket fuel, not when they began suffering more and more failures. *Salyut-6* was the first that could be refueled, and it ran for several years longer than planned until it was replaced by *Salyut-7*. *Salyut-6* even had a special 13-day visiting "maintenance mission" in which three cosmonauts made major repairs to the thermal control system and other equipment—even though it was to be abandoned within six months. The Russians were replacing the two-man *Soyuz* version with a three-man version, and they needed a new station that could host a three-man crew for long visits, so *Salyut-6* was obsolete.

Salyut-7, the last station before *Mir*, had major failures every year, but it was still operational when the last crew had to return to Earth because of a medical emergency. Nothing that resembled Blagov's "pre-avalanche" failure buildup had occurred. So I suspected that if he were waiting for this kind of ideal warning indication on *Mir*, he and his whole team might yet be caught by surprise.

However it was done, the Russians were justifiably proud of their ability to keep *Mir* operating through the 1997 crises. But they seemed to have gotten so used to dodging bullets that they developed the hubristic impression that they were perpetually bulletproof.

"There will be failures," Blagov admitted. But then he boasted, "We will overcome them as we have overcome them for the past 12 years. There have been masses of accidents since the first months, and we dealt with them. And we'll deal with them now." And in an unguarded moment of exultation after surviving another crisis later that year, the chief flight director, Vladimir Solovyov, was heard boasting, "Nothing, nothing is going to knock us off this horse."

But sometimes the teams led by Blagov and Solovyov seemed almost self-destructive in their decisions. Even after months of one crisis after another, all blamed on anything or anybody but Mission Control, they persisted in asking for trouble. For example, when the *Mir* crew went outside to inspect the Spektr module for punctures late in 1997, they were already endangered by a scheduling constraint that forced them to prepare for and perform the space walk during their normal sleep period. In other words, they had to undertake a strenuous and dangerous activity while jet-lagged. And Mission Control approved.

This risk was taken because the Russians could not depend on their robust network of relay satellites for communication. In order to speak to the crew, they had to wait until *Mir* was flying across the ground tracking sites inside Russia. And this happened to begin during the time when the crew should have been sleeping.

Such a schedule violated the safety standards for both programs, but it was necessary because of inadequate communications capabilities. Fortunately, this hazardous gamble had no serious consequences. Added risk came from Russian President Boris Yeltsin's bizarre command that the crew be awakened in the middle of their abbreviated sleep period in order to take part in a live concert celebrating Moscow's 850th birthday. Nobody at the TsUP had been able to say no to this improper political imposition. And when they got away with it, when there were no negative consequences, their risky behavior was reinforced.

8

The *Mir*
Safety Debate

*"A man should never be ashamed to own he has been
in the wrong, which is but saying in other words, that
he is wiser today than he was yesterday."*

Alexander Pope

*"Quality must be considered as embracing all factors
which contribute to reliable and safe operation. What
is needed is an atmosphere, a subtle attitude, an
uncompromising insistence on excellence, as well as a
healthy pessimism in technical matters, a pessimism
which offsets the normal human tendency to expect
that everything will come out right and that no
accident can be foreseen—and forestalled—before it
happens."*

Hyman Rickover

"Let's hope that everything that went wrong is leaving with us,"
Aleksandr Lazutkin told Mission Control shortly before he and
crew commander Vasiliy Tsibliyev left *Mir*. And Fate complied: Like a
Greek Fury, bad luck pursued them all the way back down to the ground.

After enduring six months of breakdowns and crises, the two men
boarded their *Soyuz* ferry craft on August 14 for the return to Earth.

They left Mike Foale in space with a newly arrived team of cosmonauts. They then proceeded through the normal *Soyuz* de-orbit and reentry sequence. Their braking rocket worked fine, their heat shield worked fine, and their parachute worked fine.

Then something went seriously wrong. Descending on its parachute, the *Soyuz* capsule jettisoned its heavy and now superfluous heat shield. This exposed a set of small solid-fuel motors on its bottom, along with a small radar transmitter to measure the capsule's altitude. At a height of only 20 feet, the rockets would be triggered in a half-second burst to soften the capsule's impact with the ground.

When it worked, it resulted in what cosmonauts have described as an "elevatorlike touchdown." One Russian described hearing the tall wheat in the field where he landed brushing the sides of the capsule moments before it settled gently onto the soft earth.

Not this time. Two miles above the ground, as soon as the heat shield dropped clear, the solid-fuel rockets fired prematurely. The cosmonauts felt the capsule jolt, and they realized what had happened. They knew that there would be no final soft-landing braking, and they braced for impact. The last time this had happened, in 1980, the crew on board had taken a momentary 30-G force. But it depended on the surface winds and the type of soil that they were going to hit.

"We hit the ground very hard," Tsibliyev told news reporters later on. The capsule was leaning to the right a little, and the empty right seat took a tremendous shock, while the center and left seats with the cosmonauts in them took a lesser force. "Thank God there was no one in that chair, just cargo," Tsibliyev continued. A person there might have suffered severe, even fatal injuries.

This incident was particularly frightening in light of the fierce *Mir* safety debate that was then raging in the United States. Amidst all the assurances of safety, there was the insistence that no matter what happened on *Mir*, the crew "could always escape" and land safely in the attached *Soyuz* capsule. But now even that last-ditch consolation turned out to be false.

During congressional hearings on September 18 concerning the decision to send a new astronaut to *Mir* at the end of Mike Foale's mission, the top congressional expert on Russian space systems expressed her concerns over the crash landing. Marcia Smith, a senior analyst in the Science Policy Research Division of the Congressional Research Service, testified: "The failure of the usually reliable *Soyuz* landing engines during the return of the Tsibliyev-Lazutkin crew in August is troubling both for *Mir* operations and for *ISS*. Commander Tsibliyev said that if someone had been sitting in the third seat in the *Soyuz* (which was vacant during this particular landing because of a change of plans just weeks before), he could have been seriously injured. Whether this is an isolated failure, or the result of lower quality control by the manufacturer, is an important issue."

The unanswered question was whether the *Soyuz* that had brought up the replacement crew and that at that time was hooked to *Mir* had the same flaw. If it were necessary to evacuate *Mir*, the American astronaut would be sitting in the right seat, which had suffered so much damage during the recent landing. Or perhaps the same flaw—if present—would cause a crash on the left side, or some other serious impact. Nobody knew, or seemed to care.

The crash landing should have made at least one man happy. French astronaut Leopold Eyharts had been scheduled to sit in the right seat after riding up into space a few weeks earlier with the replacement crew. Only weeks before launch, his space visit had been delayed for six months, until the next "crew exchange." The French figured that *Mir* was in no shape to host his experiments, and they wanted their money's worth. So this time the seat was empty, and the French astronaut made his flight six months later, after the *Mir* had been substantially repaired.

Months later, the Russians explained to NASA that the crash had been just another freak accident. "Their term was something like 'spurious galvanic connection,' meaning the excessive humidity may have caused a short leading to a premature activation when the heat shield was deployed," Jim van Laak e-mailed me. Van Laak was a top NASA operations manager for shuttle-*Mir*. He often disagreed with me, and I with him, but he always wanted to exchange all possible information about issues facing the project.

As for my concern that if the rockets had gone off before the heat shield had been detached, their premature firing might also have damaged the shield or the cabin itself, van Laak said that he'd been told that there were "multiple inhibits" to keep the rockets from firing before heat shield deployment. But even in the "worst case" of a firing in orbit, he explained, the rockets don't "fail the heat shield, let alone the pressure hull." That is, the flame wouldn't break through the heat shield (leading to burn-up) or the pressurized cabin's wall (leading to loss of air). As for the source of the water that had caused the short circuit, apparently it had come from the *Soyuz*'s air conditioner. Since the larger unit on *Mir* had been broken for months, the smaller one in the *Soyuz* had been turned on, and it ran well beyond its designed lifespan.

Russian safety documents that I had reviewed while working at NASA told a somewhat different story about the hazard associated with these rockets, however. I wasn't surprised to find that the document falsely claimed that the soft-landing rocket had never failed before. But I found it interesting that the hazard of a premature rocket firing was rated as "catastrophic" in the official Russian document. Vehicle and crew loss would be the likely outcome, it stated. Probably the Russians hadn't shown that page to van Laak.

In terms of discovering what could unexpectedly go wrong in space, *Mir* was certainly a gold mine. And as always, NASA did its best to put a positive spin on the sequence of shocks and unpleasant surprises.

"I learned far more in recovering from those things than if I had just had a steady constant program to execute," *Mir* veteran Michael Foale told a press conference after his flight. And White House science adviser John Gibbons, in the middle of listing the accidents, told reporters, "My guess is that if everything had gone [as] we hoped, ... we'd [have] come out on the short end of the stick in terms of relevant experience. . . . Each one of these mishaps has been absolutely invaluable in teaching us lessons that will come in very handy when we do further work in space like the space station."

More level-headed space veterans just shake their heads at such bold talk, since they know that emergencies in space are never good news and are never desirable. Further, noted one 20-year veteran of NASA's Mission Control, the claim that the crises taught NASA how to work out problems with the Russians merely shows that NASA approved the flights before it had adequately trained itself to work out such problems. "You learn to work with other control teams during rigorous pre-flight drills and emergency simulations," the veteran explained, "and this must be certified at the flight readiness review.

Nobody had any right to allow those flights if this ability hadn't been learned before launch."

Charlie Harlan had been head of the safety office at the Johnson Space Center before retiring in 1997, after a career dating back to the days of *Gemini* and *Apollo*. On June 29, 1997, he wrote a long, calmly worded letter to NASA headquarters' safety office to argue against the notion that unpleasant surprises were a good thing.

"High NASA officials and other pundits are quoted in the media as attempting to characterize the dangerous and potentially life threatening situations as 'an opportunity to learn for *ISS* and how to work effectively with the Russians,'" he wrote. "[But] in the past, we have relied on training and simulations for this kind of opportunity rather than real life emergencies of the survival category. Of course we have to deal with real life emergencies at times, and we do learn from them, but we shouldn't ever view them as an opportunity."

Harlan concluded that the lesson to be learned from the surprises was that NASA had failed in its obligation to perform adequate hazard assessments: "Those of us in the safety profession would view these events as a failure of the management system and our Safety and Mission Assurance process," he wrote. A healthy process, he added, "should be based on disciplined risk assessment methods, hazard elimination, risk mitigation, and continuous risk reduction throughout the life of the program."

In Harlan's assessment, which is shared by many of the other engineering safety experts that I talked to, the most disquieting thing about most of the problems on *Mir* was that they kept taking both countries' space experts by surprise. NASA safety panels had received detailed Russian input when they certified, before Jerry Linenger and Michael Foale's tours of duty, that nothing was likely to endanger them or threaten their missions. The Russians, with 25 years of space station living behind them, also overlooked these risks, demonstrating that their own process of extracting the correct lessons was flawed.

NASA's reaction to the 1997 accidents was mainly focused on how to prevent a recurrence. "They've firmly locked the doors behind all of the stolen horses," one former NASA safety official remarked. He meant that the two major and dozen minor accidents that actually occurred were only a small subset of all the potential accidents, thus making it wrong to believe that preventing their recurrence would protect the *ISS* against all other possible misadventures. Viewed from this angle, focusing on the history of *Mir* in 1997 and what had been learned from it could divert attention from a much wider-ranging hazard assessment.

This concern is well founded. In August 1997, astronaut Wendy Lawrence, then still a candidate for a lengthy *Mir* mission, explained to a television reporter why she was not worried by the string of space calamities. "I figure that everything that can go wrong has already gone wrong," she quipped.

Viktor Blagov, interviewed in his office at the Mission Control Center in Moscow, said much the same thing to an American news reporter later that year. "We have learnt to repair every system on board—simple or complicated, we had so many malfunctions and in so many ways that, we can say, they can only be repeated," Blagov boasted. "People are joking here—we've lived through everything in that station, nothing new can happen as we've had it all."

The delusion that nothing new could go wrong in a system as complex as *Mir* was extremely dangerous. If the Russians had come to believe that they were invulnerable, they were in even greater danger than ever. Real rocket scientists know that it isn't just a question of a single system failing. One system can cause another to fail, or a failure in one system can mask a failure in another. Those are the kinds of things that can happen in more vulnerable systems.

Harlan's letter continued: "When NASA originally began the *Shuttle/Mir* program, no rigorous safety analysis or risk analysis was accomplished." He knew this because he had been there. "NASA decided based on the then understood historical performance of safe *Mir* operations to accept that record as a given. This was done by a subjective review process unlike the systematic safety and reliability analytical techniques utilized for U.S. human space flight. If you remember, at that time the Russians were not always forthright about their systems failures or some of the problems they had in the past."

That was Charlie's diplomatic way of describing the situation that I had also complained about in much more forceful terms. Where he said that the Russians were "not always forthright," I said that they were lying. The consequences were the same: NASA accepted the false data, and neither soft chiding nor harsh criticism altered NASA's policy in the slightest.

Harlan continued: "NASA publicly has the appearance of trying to characterize the recent dramatic events of *Mir* operations in a way to minimize the idea that there is any safety concern for the crew as a result of the current *Mir* status. I believe that this stretches the limits of credibility.... Should it become clear to NASA that the safety risks for operations with *Mir* are increasing, NASA management should have the guts to challenge the political basis for this specific activity and offer alternative programs for cooperation with the Russian Space Program to the Administrator."

Within NASA, other safety experts were echoing Harlan's concerns. One was Blaine Hammond, an astronaut assigned to be the safety representative for his office. "We have been extremely lucky so far," he told his associates in July 1997. "We may not be so lucky next time and, in my personal opinion, there will be a next time, it's just a matter of when and how bad."

But when Hammond called the *Mir* "a disaster waiting to happen," he was cold-shouldered. "You'd have thought I was preaching heresy, the way people reacted to that," he told a reporter the following year, after he'd left NASA. "They would let me talk, but they didn't act like they ever were going to take it forward. You'd see eyes rolling or you'd get the impression, 'Geez, here he goes again.'"

On July 1, 1997, soon after the collision, chief astronaut Bob Cabana rebuked Hammond by e-mail for his negative statements about *Mir*. The message, Cabana explained, was to remind him of "the Flight Crew Operations Directorate position," and it ended with the blunt request, "I would like to talk with you."

"I was told that you stood up at a meeting and said, 'it would be criminal for us to send Wendy to *Mir*,'" Cabana complained. He then criticized Hammond for making such a statement without clearance. "Our primary goal right now is to help the Russians fix *Mir* and ensure that it's done correctly. Your job is to make sure the system is supporting [our decisions], doing all the right things to fly safely, not to express [emotional] personal opinions that may or may not coincide with policy."

It was clear that NASA's policy had already been set: Stick with *Mir* unless something overwhelmingly negative shows up in the studies. This was evident from an e-mail that Cabana sent to Wendy Lawrence, the astronaut then training to replace Foale on *Mir*. "I think there is always the possibility that Mike could be the last American on *Mir*," Cabana wrote on June 30. "This is definitely not what the program wants." He added that if *Mir* was found to be "safe and stable," "I'm sure we will continue."

"Not what the program wants" is a telling phrase. NASA seemed to know the answer that it wanted in advance. Despite the public pretense of an agonizing, even-handed appraisal, the agency never seriously considered *not* sending the next astronaut to *Mir*. This is demonstrated by the curious fact that it wasn't until September 24, the day before the shuttle launch to *Mir*, that shuttle operators even asked the mission design team what the dynamics issues (propelling and steering the shuttle into space) might be if the next American who was supposed to fly to *Mir* were to be taken off the shuttle before launch. The seat would have to be empty going "uphill," because it would be occupied by the returning American crew member (already on *Mir*) on the way back, so the shuttle would be slightly lighter.

James van Laak told the *New Yorker*'s Peter Maas the same thing: "To be perfectly honest," van Laak is quoted as saying about this decision, "there are plenty of people within the political system and within NASA who are pushing us to go, go, go, go, while at the same time, they are distancing themselves from any blame."

Why wasn't the issue being studied seriously? "The problem is lack of communication up and down the line," Hammond later explained. "Nobody wanted to hear me, I felt I was left out in the cold," he later told the *New York Times*. And it was more than just not wanting to hear things. Apparently there was also some effort expended to make sure that others didn't hear things, either.

Hammond recalled a lively discussion of *Mir* dangers during a teleconference between Houston and NASA headquarters. The meetings, like all formal procedures, were taped. When a truly independent oversight team—NASA's own Inspector General's office—heard about these meetings, it requested the tapes. Johnson Space Center officials reportedly replied that a person or persons unknown had broken into the Building 8 vault where the tapes were kept and that unfortunately these 17 specific tapes were missing.

The suspicions of the skeptics—Hammond, Harlan, myself, and a number of other holdouts—were that NASA was assuming that *Mir* was safe until proven otherwise because that was what the agency wanted to believe. This was the classic sin that led to the *Challenger* disaster in 1986, when a manager demanded that engineer Roger Boisjoly "prove it isn't safe" when Boisjoly correctly voiced concern that the engine seals hadn't been verified at the low temperatures they were facing on the day of the launch. It was the same philosophical flaw that appeared in 1999 when NASA managers, informed that their interplanetary navigation experts were worried that there was something funny about the flight path of a Mars-bound probe, demanded that the navigators *prove* that the probe was off course before they would allow the requested slight course change that would lead the probe to pass a little farther above the planet's atmosphere. Without the prudent adjustment, the navigational problem, which turned out to be all too real, caused the probe to literally crash and burn. In public, NASA explained that debacle as a mixup of metric and English units, instead of the true failure of management judgment.

Written proof of the same philosophical flaw was found in the briefing charts of a NASA safety panel that met on September 10, 1997, to consider the launch of the next astronaut to *Mir*. Pete Rutledge, executive secretary, presented the chart that spelled out the panel's position. "Despite concerns," the chart said, "there is no hard evidence that *Mir* is currently unsafe."

There it was in black and white. NASA's view was that it wasn't a question of showing that the agency understood the potential hazards of *Mir* and was taking trustworthy countermeasures. No, NASA's arguments were based on willful ignorance. NASA felt that since it couldn't find any dangers, it should assume that there weren't any. This was the agency's position even though the same review process had been used before Linenger's flight and before Foale's flight, and the process had obviously failed.

Rutledge's recommendations took diplomatic considerations one step further into what should have been a straightforward engineering assessment. "If and when *Mir* is deemed unsafe for a U.S. presence," the chart said, "NASA should convince our Russian partners that *Mir* is unsafe for a continued human presence and press for abandoning the vehicle completely." It was a macho thing: If the Russians wouldn't quit, we wouldn't either.

It's worth making the point again. Normal space flight safety standards, revalidated after the *Challenger* catastrophe in 1986, call for establishing a positive level of safety by cataloguing all of the potential hazards and estimating their cumulative probability in the face of countermeasures. In contrast, NASA's approach to *Mir*, as illustrated in both internal briefing charts and explicit public statements, was to assume safety unless danger could be proved (the motto seemed to be, "We will proceed unless somebody proves there is a reason to stop"). But most of the dangers facing the astronauts on *Mir* were unknown. Either Russia withheld relevant information, or else no one checked up on *Mir*'s key features. A full year later, in June 1998, a senior Russian space engineer was brought up to *Mir* on the last shuttle docking with one main task: to check *Mir*'s soundness. Nobody had had any reliable information about it before.

NASA's logic as described by NASA spokeswoman Peggy Wilhide was upside down, challenging doubters to find reasons *not* to go. "The bottom line was that the experts that we had asked, the majority of them, determined that there were no technical or safety reasons to discontinue the program," she told the Associated Press.

In fact, could the safety teams be expected to find reasons *not* to go? In order to perform his safety assessments, Apollo astronaut Thomas Stafford (who was heading a special independent review panel) relied on U.S. consultants and on his opposite number in Russia, a space engineering veteran named Professor Vladimir Utkin. Utkin arranged briefings for Stafford's people in Moscow and assured them that he could find nothing hazardous about *Mir*.

This concerned me because Utkin's track record in assessing space hardware hazards didn't strike me as encouraging. A year earlier, he had been in charge of the accident investigation for the *Mars-96* probe. The probe had crashed to Earth shortly after launch in December 1996, scattering half a dozen plutonium-powered batteries across the Chilean-Bolivian border. Utkin's panel was given full technical data on the probe and its booster, yet even knowing as they did that it had actually failed, they could find no technical reason for it to have done so. Had they been asked prior to launch to predict the outcome of the mission, they would have concluded that they could find no reason for concern, even though, in the real world, the launching turned out to be a disaster.

Nor, apparently, were many of NASA's consultants given adequate technical information. Dr. Ronald Merrill's report on the 1997 fire is a good example of this. Commenting on the fire, Merrill stated that "the smoke seen at the time of the fire proved to be water vapor and not a serious environmental issue." I compared this view, based only on some documents that Stafford's group had selected to show him, with Jerry Linenger's written report two days after the event: "Breathing not possible without gas mask," he wrote. "In my judgment, survivability without gas mask—questionable; serious lung damage without gas mask—certain." Concluded Linenger, "It was impossible to get even a single breath between masks due to smoke density. I have experienced similarly dense smoke only during Navy firefighting training." Linenger couldn't see the fingers on the end of his hand through the smoke. I found it impossible to believe that Merrill had been shown Linenger's full reports.

Other private analysts found different weaknesses in Stafford's conclusions. "The full Stafford report . . . appears only to address past safety problems and not root causes of why safety issues occurred or how Russian flight safety assessments will be different in the future," wrote Dennis Newkirk, author of the *Almanac of Soviet Manned Space Flight* and editor of the Internet's *Russian Aerospace Guide*. "The summary of the report also neglects cascading failures in which one systems failure, such as cooling, can eliminate multiple other systems leaving *Mir* with no backup other than the *Soyuz*."

At the hearing, Roberta Gross, NASA's Inspector General, described concerns expressed to her office about the impartiality (or lack of it) of the review boards. She recounted that she had received numerous comments from space workers to the effect that the review boards, which consisted of people whose real jobs were dependent on NASA funding, could not be impartial. The U.S.-Russian panel headed by former astronaut Tom Stafford was specifically mentioned.

Ralph Hall, a Texas congressman, attacked Gross for relying on "anonymous hearsay." He was clearly overlooking the fact that the people who spoke to Gross's team gave their names, but on condition that their identities not be revealed to their management (that's how an effective Inspector General works in any bureaucracy). "I find many aspects of this testimony very troubling," he complained, citing the practice of repeating anonymous charges without Stafford being present to respond (he had been invited but had a scheduling conflict). "That's an indictment of General Stafford and George Abbey," Hall thundered. Such charges "smear the reputation of Tom Stafford," he complained, and he added, "Making allegations that question his integrity—I think that's disgraceful." Hall had been assured by NASA that the space station would discover a cure for cancer, and he was a "true believer."

But Gross's reports clearly portrayed an atmosphere within NASA in which workers knew that disagreement was not tolerated. One non-*Mir* investigation examined a controversial decision in Houston to eliminate all Mac desktops and switch to an all-PC environment. "One employee stated that [s]he was told by the supervisor, 'I would hate to lose a good engineer over this,'" Gross wrote. "Another employee was told their value to the organization would be seriously questioned if they continued to question the decision. In another instance, a spokesperson sent an e-mail stating 'Our management is getting very tired of people who always know better. I know who signs my paycheck, do you?'"

Further indications that this feeling was widespread came from a discussion forum on the private *NASA Watch* Internet site. The topic was, "Do you feel free to openly express your thoughts at NASA without fear of retribution?" Some said they did, although few of those were from Houston. Otherwise, there was a lamentable consistency in the replies.

"In a healthy group, disagreement is not a threat," wrote one. "Rather, disagreement is part of a process that results in the best decisions. Leadership assures things get done, even if disagreements persist after a decision. The problem in the agency is too many managers and no leaders to be found. Managers squelch disagreement. They take the easier route than a leader would. Somehow we got top heavy on managers and scarce on leaders."

Another engineer expressed this concern: "I fear that the 'yes' men [and women] will say 'yes' one too many times, and there will be big problems that surface."

"You know very well there's no way to do that in this organization," wrote a third. "We've become a parody of the worst of [Marshall Space Center, which supervised the shuttle propulsion system] just before

Challenger—a bunch of 'yes' men without the guts to tell the emperor he isn't wearing any clothes. Under this administration and this Administrator, NASA has become an agency of lies and half-truths, especially with regards to safety."

Another wrote: "Do I feel free to speak out? Not a chance in hell! I've personally seen what happens to those that do. Around here we call such behavior 'career limiting moves.'" And another: "Many NASA and contractor employees have been in positions to see serious management mistakes, waste, breaches of the public's trust or just plain corruption, but have not had the courage to speak up." And yet another: "Typically, it is the employees that are of retirement age and who have no further career aspirations that are less apprehensive about speaking their minds. If you don't fall into this category then I would say you probably are a little more reserved in what you say. 'Rocking the boat' and not 'towing the party line' can and do end careers. I've seen this happen more than once. We are creating 'yes-people.' This is unfortunate."

Goldin, the head of NASA, was aware of these concerns, and regularly made announcements stating that the problems should be fixed. In mid-1998, for example, the Senior Staff Meeting Minutes for June 1, 1998, relate that "Mr. Goldin expressed his concerns that NASA does not have an environment that is conducive to openness in addressing safety issues raised by employees." He was quoted directly: "Open communications and employee awareness and training are critical to produce an environment that permits information and concerns to flow and be appropriately addressed." Somehow it never happened. Two years later, taking blame for the latest NASA failures with Mars, he confessed that "people were talking, but we weren't listening." He still didn't get it: People weren't talking because they were intimidated when they tried to pass bad news up the management chain.

Although NASA declared that the decision about whether or not to send the next American to *Mir* must be made unemotionally, there were transparent attempts to depict the worriers as cowards. Frank Culbertson told the *Washington Post*, "I'm worried about whether the country still has the courage to do high-technology projects like this." His opposite number in Russia, Valery Ryumin, was equally blunt. "I think it would be a very unwise decision" to curtail the flights, he said. "You would have a situation where it would appear the Americans are 'sunshine space explorers' where as soon as something goes wrong, they head for the hills."

Some news commentators agreed. "The general American public—its voice reflected in its heroes, politicians and late night comedians—is no longer willing to accept risk that is inherent in spaceflight," wrote Jim Banke, manager of the Space On-Line press service in Florida. "Apparently we dream of exploring distant shores but won't swim far into the ocean when the saltwater stings our eyes." Speaking from my own perspective, I found that kind of talk insulting because I knew that men such as Blaine Hammond were taking far greater risks than those who were telling the management what they wanted to hear.

NASA did appreciate the public relations challenge that it faced in justifying its decision. A "*Mir* Safety Assessment Briefing" given to top managers by Frank Culbertson specifically addressed the need to "educate" the public and government leaders about "the perseverance required for significant exploration efforts." The success of this process would depend "on the ability of public and government to understand the realities" of space exploration. Culbertson warned that "abandoning the Phase 1 [the shuttle-*Mir* visits] project sends the wrong message."

The Russians made it clear to NASA that the entire *ISS* program might be at stake if NASA pulled out of Phase 1. "If operations with *Mir* are discontinued," warned Valeriy Ryumin of Energia in another internal memo I obtained, "this will not speed up the construction of Alpha [Phase 1]." Without making any explicit threats, Ryumin made sure that the American side knew the stakes: "On the contrary, this could result in its construction being stopped entirely."

I had been invited to testify at the congressional hearings on September 18, to discuss my independent assessments of NASA's safety review process.

I told the committee that while NASA officials kept insisting that "*Mir* was safe," they had no engineering justification for making such an assertion. This was because the familiar process of ground-up safety assessment, which had worked so well in the past, had never been applied in this case. In order to objectively prove that something is safe, it is not enough to challenge others to "prove that it's *not* safe," all the while withholding pertinent information, and then triumphantly conclude that the absence of any proof of danger is equivalent to a proof of safety. But that's how NASA had done its *Mir* analysis.

I mentioned one minor example of information that was potentially related to safety that did not get to the people who were supposed to have it—that is, the safety workers and the public at large. A few weeks after the collision, *Mir* reported that there were "popping" sounds coming from inside of the abandoned *Spektr* module. They also spotted "snow" jetting out from its side. These observations were overheard by Dutch radio listener Chris van den Berg, but neither NASA nor the Russians disclosed them until they were asked about them by a journalist who had read an Internet report by van den Berg. After analysis, the noises and the flakes were explained as exploding shampoo bottles.

At the hearing, I was asked if a new astronaut should be sent up to *Mir*. I replied that six months from now, the question of leaving Americans aboard *Mir* could be raised again. If repairs to *Mir* systems had by then proven their efficacy, if redundancy had been restored to all life support systems, if nothing else major had broken down, and if the *Soyuz* landing accident investigation results were credible and supportive, that might well be the time to consider renewed long-term American visits. The answer at that point might be yes, but at the date of the hearings, I urged that the answer be no.

Marcia Smith from the Congressional Research Service was the last to testify. As always, she had done her homework with precision. And as always, she presented her devastatingly unarguable observations in a calm, noninflammatory, and precise manner.

"NASA also argues that the United States needs to fulfill its obligations to its partners and not be a 'fair weather friend,'" she stated. "In this context, it is important to remember that the last two shuttle-*Mir* dockings (scheduled for January and May) and most of the time accounted for by the last two astronaut visits to *Mir* were agreed to in 1996" as a supplement to the original series of visits. Without them, the shuttle mission in September would have been the final flight in that sequence, and no further Americans would be left on *Mir*.

What she was reminding NASA was that the need to decide whether to send another American to *Mir* in September 1997 would never have come up if NASA, two years earlier, hadn't expanded the shuttle-*Mir* program. Without that decision, the September return of

Mike Foale would have been the end of the U.S. presence on *Mir*, independent of any hazards or other considerations.

Smith disclosed the nature of the deal that had added the two shuttle flights. "In exchange, Russia made three promises—to keep the *FGB* module and the *Service Module* on schedule, and to build a new cargo version of the *FGB*," she explained. Under that plan, NASA agreed to make additional payments to the Russian space agency, nominally as "rent" for being allowed aboard *Mir* longer than originally contracted for.

"Only one of these [promises] has been kept," she pointed out. "The Russians could have delivered the *FGB* on its original schedule, though the schedule slipped because they could not meet the *Service Module*'s schedule."

In strict deadpan, she concluded, "This demonstrates that there is flexibility in meeting partner obligations." This trumped NASA's argument that the United States was obliged to send up a new *Mir* visitor in order to show that it was a reliable partner, even though the promises Russia made in order to arrange for this new visitor had been broken because Russia had turned out *not* to be a reliable partner. NASA was insisting that the United States remain bound by a deal even though Russia had already broken its half of the bargain.

I would have said it differently, but probably not as effectively. Continuing with the supplemental *Mir* missions after the Russians had already failed to live up to their side of the deal would teach them a very important lesson: There would be no penalty, none at all, for breaking promises to NASA. And Russia, as it turned out, would quickly become used to getting away with just that.

9

Rescue and Recovery

*"If any man can convince me and bring home to me
that I do not think or act aright, gladly will I change;
for I search after truth, by which man never yet was
harmed. But he is harmed who abideth on still in his
deception and ignorance."*
Marcus Aurelius (121−180) *Meditations, vi. 21*

*"Mishaps are like knives that either serve us or cut us
as we grasp them by the blade or the handle."*
James Russell Lowell

When Anatoliy Solovyov arrived aboard *Mir*, it was as if a switch had been thrown. An almost electric sharpness instantaneously replaced the fatigued, dispirited attitude of the station's crew members. It was August 7, 1997, and the 48-year-old Russian cosmonaut exuded the confidence and determination necessary to end *Mir*'s long string of crises.

"This was 'Captain America,' or 'Captain Russia' if you will, coming aboard to save the day," Mike Foale later confided to some of his associates back on Earth. "He and Pasha dove into the recovery operations from the very first moment they arrived."

Solovyov hadn't even been supposed to return to *Mir*. Six months earlier, as Russia's most experienced space pilot, he had been named to the first crew for the *International Space Station*. But an American astronaut, Bill Shepherd, was designated mission commander. Solovyov would merely have been the *Soyuz* commander during the flight up.

This didn't sit well with Solovyov, a veteran cosmonaut who had been in command of every one of his four previous space missions. He had racked up more than 450 days of space experience, docked six times to *Mir*, performed nine space walks with a record-setting duration, even been launched aboard a NASA shuttle once. He had performed both normal long-term operations and emergency repairs. Once, he and a crewmate had been trapped in outer space when their airlock hatch malfunctioned, but they had been able to use a backup airlock to get back inside.

Understandably, Solovyov refused to serve under the command of a much less experienced American. The prospect of years of launch delays also probably didn't go down well with the senior cosmonaut. Program officials sympathized, and they allowed him to swap assignments with the commander of the next *Mir* expedition, Yuriy Gidzenko. Although Gidzenko looked to be about 20 years old, the highly intelligent 35-year-old pilot had already spent half a year in space. He'd be 38 before he flew again on his new space assignment.

For his partner on *Mir*, Solovyov had Pavel Vinogradov, a civilian engineer who had also built up a serious determination to get into space and triumph there. The previous year, he had been within a week of being launched to *Mir* when his crew commander failed a preflight physical and the backup crew took over the mission. At 44, he was already a little old to be a space rookie.

Perhaps the timing was only coincidental, but a new tradition was inaugurated on this flight. At the cosmonaut hotel, as the crew signed their room's door and performed other cosmonautical preflight rituals, a black-robed Russian Orthodox priest gave them a blessing for travelers. If any space mission needed prayer, it was Solovyov's.

On the way to the launch site, by tradition, the bus stopped so that the two men could get out and urinate on the tire. Gagarin started this ceremony unintentionally in 1961. Subsequent spacefarers, in a tribute to the superstitious impulses that humanize the awesome technologies that they stake their lives on, didn't want to tempt "bad luck." Female cosmonauts even carry a small flask of their own urine so that they can make the obligatory yellow stain. In 40 years, the only space-bound crew member who had declined to do this was Norm Thagard; he was worried about TV cameras with telephoto lenses.

Six weeks after the Russian crews had been exchanged via *Soyuz* flights, the NASA crew member was swapped on a shuttle flight. Mike Foale went home, and Dave Wolf arrived. (He had replaced Wendy Lawrence when it was found that her arms were too short to handle the Russian space suit.) His first impressions of *Mir* were not favorable. "It was a filthy bachelor pad when I got there," he later told me. "Food crumbs and coffee stains were everywhere. Condensed water clung in globs around equipment and on walls behind panels."

It may have looked bad, but it didn't smell that way. "There was *no* smell," he recalled. "Their filters really remove all odors very effectively. I remember my sense of smell sharpening." This sharpened sense persisted for his entire mission, leading in the end to another major sensory surprise: "When we landed and opened the shuttle hatch, I was assaulted by odors of grass, of bushes, of cars, even of buildings. I thought, 'This is how a dog senses the world.' After a few days it returned to normal."

Wolf didn't need any heightened sensitivity to perceive his assigned role on the station, as determined by Solovyov. "He was an intensely nationalistic Russian, and he wasn't all that sure that our partnership was the right direction to be going. He doubted the value of the American involvement; he was sure the Russians could do everything better in space."

This didn't bother Wolf. As a doctor and a colonel in the Air Force reserve, where he had been a flight surgeon, Wolf was used to being a good team player. He later told me that this military experience was important to his success on *Mir*. "A military background takes the 'self-ness' out, the idea of doing things 'for me', where each task is seen as a way of getting 'me' ahead. No, the important thing is how to get the job done. It taught me to jump in and do the hardest, most difficult task. I was able to feel good doing the junior guy's duties in flight. That kind of teamwork is the kind that the military clearly instills."

Regarding Solovyov's attitudes, he told me, "I think the mission did a lot to change his mind. My goal was to make them not able to get along without my help. I took on all the dirty work that they couldn't." During his flight, he had provided more details via e-mail: "I've taken on

as a particular project to keep the air filters clean. All the dust and parti-
cles float through the air and air filters clog up very quickly and it's a lot
of work to keep them clean."

Solovyov made sure that Wolf was following his physical training,
staying fit both for the eventual return to Earth and for the physically
challenging space walk that they would make together. "He constantly
pushed me to exercise harder in space, particularly to strengthen my
legs and back," Wolf recalled. "It helped me to have a coach. Everything I
did, he said, 'That's not hard enough.'"

Wolf's previous space mission had been STS-58, for 14 days in
October 1993. The shuttle carried a *Spacelab* science module loaded
with life sciences experiments and 48 rodents. Two other members of
that crew, John Blaha and Shannon Lucid, had later also flown to *Mir*,
and Wolf had studied their experience.

Short flights and long flights are two very different kinds of space
missions, he knew. "The shuttle flight is a sprint race," he explained to
me after his *Mir* mission. "It's highly choreographed." The energy
budget for shuttle missions and space stations is also different. "You can
dip into your pre-flight energy reserves to some degree," Wolf said of
the shuttle flights. "On *Mir*, you have to operate at a level that can be
sustained indefinitely."

He thought for a moment. "But you know," he continued, "on my
easiest day on *Mir*, I still worked harder than on my hardest day on the
shuttle." An explanation occurred to him. "On shuttle flights the crew
activity is constrained by planning guidelines," he suggested. "I guess
they need to hold back some reserves for emergencies." It was different
on *Mir*. "Since you don't have to land immediately, you can coast, handle
the emergencies, rest in between," he explained. "On the station, you
really can run further into your reserves."

But the Russian station also provided Wolf with inspiration. "My
cubicle (really the airlock) has a view that is out of this world. I share it
with three space suits," Wolf e-mailed on October 6. I poetically replied
that I was viewing the module pointing upward like a tower, the rest of
the station as its lower chambers and cellars, with his abode a slim,
white, flat-roofed wizard's tower with a four-way view out into the universe, on the rim of the known world, with a magic door to step out
through, and golems to guard his conjurings. What a place to dream in!

And dream he did, he later answered, although his dreams were mostly standard Earthside dreams. The only thing that had changed was that everybody was floating, not walking.

Wolf slept naked in his sleeping bag, except for his "crash kit" in a waist pack. And he knew his way from his sleep station to the *Soyuz* escape capsule in the dark. Prepared for the worst, he slept very well.

"For power failures, I had extra batteries, maybe ten AA's, plus a little food and drink, a pocket knife," he explained to me. "I had a small bag of exposed film, computer disks, a few other items. I had pre-positioned some warm clothes in the *Soyuz*."

He practiced the run several times, and figured he could make it in less than a minute from the time an alarm woke him. "I figured I'd get to the hatch and enter the *Soyuz*, and look back and see if the other two guys were coming," he explained.

Wolf took another item with him into space that gave him, he told me, additional reassurance. It was my private list of concerns, the unverified safety-related issues that I felt had not been adequately checked out before NASA decided to send him into space late in 1997, despite the series of near-fatal accidents on *Mir*. He intended to check them out personally.

One concern of mine had been the air pressure integrity. I was worried about air leakage and the possibility of a sudden catastrophic failure of a hatch or even of a panel of the station's aluminum hull. This had happened dramatically on an Aloha Airlines flight a few years before. There, aluminum skin that had been repeatedly flexed had developed undetected microcracks that had led to a sudden blowout of much of the cabin roof during flight. Although one crew member was killed, the pilot was able to land safely, saving everyone else on board. That would not be the result if a similar panel blew out on *Mir*. With all of the air gone in a matter of seconds, everyone would die before they could reach the lifeboat.

During the congressional hearing before Wolf's launch, Frank Culbertson testified that his greatest concern was that *Mir* would cease to be airtight. But according to his testimony, the only possible causes for such a failure were meteorites or collisions with space debris. That was exactly the way it had been specified in the "*Mir* Safety Assessment Briefing" that NASA had held the week before. The chart that read

"Known or Potential Hazards" included "Depressurization," but it listed only a single conceivable cause: "Collision."

There were at least half a dozen other ways in which air could have leaked out of *Mir*, and once again, NASA had simply overlooked them. The seals around the airlock hatches could fail, or the hatch mechanism could fail. This was particularly worrisome because the hatch opened outward, so that the internal air pressure was always fighting against the hatch, not pushing it more tightly into place. The hull structure could crack under mechanical or chemical stress. It was worrisome that *Mir*'s aluminum hull had been doused in leaking ethylene glycol coolant for more than a year. On Earth, this coolant is explicitly contraindicated for aluminum-block automotive engines because of the likelihood of corrosive reactions. Behind *Mir*'s consoles, numerous pipes connect interior instruments with the outside vacuum, and valve command failures or piping breaks could create another conduit from inside the station out into space. Each of these hazards can be controlled by safety procedures. At one time or another, each of these hazards had caused air leaks on *Mir* or on earlier Russian stations. But NASA refused to acknowledge that any of them even existed.

Metal fatigue had been building up on *Mir* throughout the repeated thermal cycles of 16 sunrises and sunsets per day. The structure was flexed when it was twisted around by steering jets. It also expanded and contracted as air was added and then lost on each space walk.

One of Wolf's American predecessors on *Mir* had told me privately about an alarming observation that he had made earlier that year. Four heavy modules extended outward from a crossroads segment on the front of *Mir*'s "base block" section. My astronaut source had been taken to the outer end of one of these modules, from which he was able to look back out the hatch into the crossroads segment and then through the hatch of the opposite radial module, all the way to its far end about 70 feet away. His guide then began to repeatedly throw himself against the wall of the section that they were occupying, in a carefully timed sequence.

"Look down the tunnel," he was instructed. When he did, he saw the far end of the opposite module begin to sway up and down. Within seconds, it was swaying so much that it moved out of view, as seen through the hatches. "It was like looking down a subway train going through a curve," he later told me. But unlike with a subway train, in which the cars are connected with hinged couplings, on *Mir* it was the metal skin of the neck of the modules that was flexing back and forth.

The pressure hull is 2 millimeters (mm) thick on average. In the node it's 5 mm, and in the small-diameter module work compartments, it's 1.2 mm. In October 1983, aboard the *Salyut-7* station, corrosion was detected on a section of the hull that was only 1.8 mm thick. "There arose concerns for the durability of the hull, for the pressurization of the station, and for the safety of the crew," a Russian report stated. The Russians finally decided to fabricate a flush-fitting metal patch that was sent up on a supply ship and glued over the corroded area.

NASA had heard about corrosion on *Mir's* inner hull, and it had followed up on the reports. In May 1997, two years after astronauts first began to stay on *Mir*, an astronaut on a visiting shuttle flight was specifically asked to assess the problem. "Most of the comments about corrosion were actually misidentified grease and other dirt," Jim van Laak later explained to me. "I was completely fooled on several of the pictures until Charlie [Precourt] told me what I was really looking at." I later asked NASA for details on Precourt's observations—say, a written report or briefing charts—but the agency declined to provide me with anything.

A year later, however, top Russian spacecraft experts remained concerned. Valeriy Ryumin, an experienced Salyut space station cosmonaut who had been given a seat on the last shuttle mission to *Mir*, described his task at the preflight press conference: "I will be involved in addressing a list of questions compiled by our station designers, questions that only a person who is experienced in this field can properly answer," Ryumin explained through an interpreter. "These questions include issues of the condition of the hull, the condition of the [power] cabling, and the conditions of various feed-throughs between modules." He certainly didn't sound as if he was unconcerned about the corrosion question.

Wolf, too, needed to satisfy himself. When I met with him after the mission, he told me that he would grab spare hours here and there to search the inside surface of the outer hull for signs of weakness. The quick answer was that he didn't find any, and this brought him tremendous relief. "There was a thin white patina on the aluminum," he later told me. "It looked like an eggshell." But the layer only coated the still-solid skin. The hull remained pristine in many areas; it was slightly corroded in others, but there was no sign of surface cracking.

Even getting close enough to the hull to inspect it was a major challenge. Wolf had to open access panels, swing out the regular equipment that lay behind them, and then dig through a decade's accumulation of spare parts and surplus equipment that generations of cosmonauts had been stashing in their "space attic."

The space between the equipment racks and the outer hull was supposed to act like a plenum, allowing the flow of warm air along the hull to keep it warm even when it was facing away from the sun. But as the spaces behind the panels became cluttered, the air flow slowed or was diverted. The outer skin on the dark side of the station cooled, and humidity from the air began condensing in sheets that were held against the skin by surface tension.

"The layer of water could be 3 inches thick in places," one American astronaut later recalled. Sometimes it was in layers, sometimes it was in grapefruit- and even basketball-sized spheres. "And it wasn't just water, it had picked up dirt and lubricant and other impurities until it was a disgusting soup that just swayed back and forth in my flashlight beam. It was a real swamp." Dave Wolf called it "slime," as in the *Ghostbusters* movies.

The water was only one item on the long list of problems that Solovyov's team faced. They tackled these problems one by one and fielded the new ones that kept showing up. Solovyov did this for six months straight, at first with Mike Foale, then for four months with Dave Wolf, and for the last few weeks with Wolf's replacement, Andy Thomas. When he got back to Earth, he described himself as being "truly tired," and he admitted that his fifth time in space was "probably my tensest mission."

But Solovyov took immense satisfaction in "rescuing" the station. "The main thing is that the station, whose condition was far from ideal, works to a full capacity again," he told news reporters. "Literally before our eyes the station began to work, modules got warm, moisture disappeared. Naturally, I as a Russian believe in our hardware and praise it, but all of the American astronauts who had come to *Mir* were also astounded by the excellent condition of the station."

Solovyov described how he kept astonishing visiting Americans with the Russian ability to repair just about anything. "When on December 31 we assembled the carbon dioxide absorption system

Vozdukh," he recalled, "American Dave Wolf looked at all this 'square eyed,' because it beat him how such a system could be assembled in space, and so that it works."

He said that the Americans who came to *Mir*, unlike the Russians, "absolutely don't understand what a long mission is and are cautious about our hardware." However, he said, the Americans get the hang of it in a few weeks, they "go through a good school," and they are new people by the time they leave for Earth. "There is a completely different feeling when they leave."

Wolf's own accounts confirm this. He dove into his tasks with an eagerness that impressed and pleased the Russians. And he quickly adopted two of their standard practices: to be willing to fix anything, and to be sparing with what you tell Earth that might worry them.

During his own debriefings back on Earth, Wolf casually mentioned that since he was an electrical engineer by training, he had taken on most of the soldering duties on the station. In a memo from the *ISS* safety team, an engineer who had heard this disclosure wrote with alarm: "Please note that we didn't know of *any* soldering going on up there." And then Wolf mentioned that he had performed maintenance on a number of items—for example, his shaver, which had shorted out in flight. "This raised some payload safety engineer eyebrows," the memo continued, "since we hadn't heard of that" either.

"The Russians have mastered craftsmanship with all the tools, the hacksaws, drills, and wires," Wolf later told *Space News*. "This is a whole level of skill that is critical to the long-term maintenance and upgrading of a space station."

Russian tinkering skills are not the only important factor here, since Americans are also known for their mechanical aptitude. The important theme for future space ventures is that Russian equipment seems particularly well suited to in-flight repair, cannibalization, or modification.

A wonderful example of this cultural trait was shared with me by Dan Roam, a friend of mine who ran a computerized office in Moscow for several years. It described an event that occurred in his office in Moscow, but it could have just as easily been describing what was happening on *Mir*.

"Because the technology and construction of machines was always sensible but bare-bones, Russians have always been obliged to know how to repair on the fly," he wrote me. "Any Russian driver knows exactly how to make a temporary fix on his Lada/Zhiguli model #7 [car] to get home after something major goes wrong (which it does at least twice a month).

"My favorite example occurred here [in 1991]. My company was among the first to use Macintosh computers in the office. One day my 21" 32bit monitor simply fizzled out. What to do? No authorized Apple dealers/service centers, no tools, no access to parts.

"Our company driver (a former electrical engineer) brought in his amp tester, completely disassembled the monitor (sending shivers down my spine when he blatantly disregarded the 'opening this panel voids any guarantee' sticker on the back) and proceeded to test each soldered transistor.

"After about ten minutes he handed me a 5mm length capacitor labeled 'Philips X2345 2 amp' which obviously had burned out. 'No problem,' he said. He called his old shop and told them what he needed. An hour later we had the Soviet version of the same capacitor. It was about the size of a salt shaker. He soldered it into place and—the monitor worked. We cut a hole in the plastic case to make room for the 'non-native' part and put it all back together with liberal use of electrical tape. To this day the monitor is a working example of Russian-American relations."

Ultimately, more than a hundred different people boarded *Mir*, and actually there were more Americans than Russians. Most of the Americans were visiting shuttle crew members, and they only stayed for a few hours. Most of the Russians stayed for months, one of them for 14 months. Sergey Avdeyev, on three trips, racked up more than two years on board.

Anatoliy Solovyov holds the record for number of visits, at five. Another Russian made four visits, and two Russians made three. One American, Charlie Precourt, also made three separate visits to *Mir*, and three Americans went there twice.

Probably the last man to expect to be aboard *Mir* was Andy Thomas. Before leaving for Russia, where he was supposed to serve as the backup man for the last American visitor, he told me over lunch in the NASA cafeteria that he saw *Mir* mainly as a "learning experience" in preparation for the *ISS*. Besides, he noted after his first space shuttle mission, there was a long line of younger astronauts ahead of him for a return flight to space aboard a space shuttle. So he chose to get into the much shorter line over in Russia.

Then, through a series of unexpected eliminations, the line short-ened even further. One candidate was too tall, so out he went. Wendy Lawrence's arms were too short for the Russian space suit, so out she went. Other potential candidates were already busy with more rewarding proj-ects and declined to compete. With just four months warning, Thomas learned that he would be flying to *Mir*, and in January 1998, he arrived.

One thing that he omitted in his rushed preparations was learning how to coordinate his stories with those of management. He often made truthful statements that contradicted the official statements. For exam-ple, after a month in space, he readily admitted, "Yeah, I would say it's undeniably tougher than I expected. It is a big challenge to do this." But in a NASA interview with Frank Culbertson, released the next day, Culbertson told a different tale: "Andy's definitely working hard and it def-initely is a challenge for him. I believe he had a good understanding of what it was going to be like before he started, and he's confirmed much of that."

As Thomas's mission drew to a close, he found that he had tempted the Fates of Space once too often. "It's true that in the previ-ous year there were a number of problems on *Mir*," he wrote home on May 29, "but I think if you look at the more recent history you'll see that the situation here has been remarkably stabilized, and we've got a very benign operation right now." He concluded, "Everything is in very good working order."

The next day, as if in response to the taunt, *Mir*'s main computer failed again, sending the station into yet another emergency power-down and slow tumble. Rehearsing their well-known parts, the crew entered their *Soyuz*, powered it up, and used its jets to turn *Mir* back toward the Sun. "We've seen this several times before," a NASA spokesperson said with resignation. "There doesn't seem to be a whole lot of alarm out there."

After commands from Earth reactivated the computer, it failed yet again. This time, however, it indicated a specific fault in the hardware. The malfunctioning unit was then replaced with another that had been brought up on the previous shuttle mission. However, the replacement also failed, and for some time, the shuttle launch to bring Thomas back to Earth was threatened with delay.

An interesting feature of the last shuttle mission to *Mir* was who was on it and who was not on it. My awareness of this started with a cryptic e-mail that I had received from a deep-throat NASA source a few months earlier: "Boy, do those Russian bureaucrats love to fly on the shuttle!"

Two Russians had flown on shuttle missions in 1994–1995 as part of an exchange for Norm Thagard's flight to *Mir*. But in mid-1997, Russian crew members again began appearing on shuttle flights, and I was able to determine that there really hadn't been any plan for this. It had all been arranged between individuals conducting their own "foreign policy."

The first to go was *Mir* veteran cosmonaut Elena Kondakova, a talented engineer who was also the wife of Valeriy Ryumin, a former cosmonaut and top RSC-Energia official. As far as I could tell, she was first invited to be a member of the STS-86 crew by the commander, Charlie Precourt, after he met her in Russia. NASA officials must have recognized the political advantages of this and approved the addition, even though she took what would have been a NASA astronaut's seat.

The next shuttle mission to *Mir* was commanded by astronaut James Wetherbee, a gangly Navy pilot who was a particular favorite of Johnson Space Center director George Abbey. Wetherbee had flown the first shuttle rendezvous with *Mir* in early 1995, but he hadn't docked, and the Russian cosmonaut Vladimir Titov had also been on board. A friendship developed between the two men, and Wetherbee invited Titov along on his next flight, *STS-86*, in September 1997. "Interesting part of the equation is that the US side pushed to have him on the Shuttle by name," a NASA friend e-mailed me. "The Russians did not want to send anyone, but we twisted their arms." With Abbey's backing, Wetherbee got his way.

Two more shuttle dockings to *Mir* remained, and the Russian space agency cast its eyes on the seats. The first went to a rookie pilot named Salizhan Sharipov, and the second was claimed by Valeriy Ryumin. Ryumin was nearly 60 at the time and hadn't trained for space for almost 20 years. He was seriously overweight, he spoke no English, and he was a chain smoker. But when NASA offered the seats to "veteran cosmonauts," he got himself picked. (As the Russian director of the shuttle-*Mir* program, Ryumin had all the authority that he needed to select himself.)

"I admit anybody else would be able to do this task as well," he told reporters a few months before his mission. "But I was the one to come up with this proposal to my management and this proposal was approved." Later on, he added another justification: "I think I have a right to see what happened to something that I spent all my life developing.

"I've come to a very interesting conclusion," Ryumin continued, describing his assessment of *Mir*'s condition. "The older a space station is, the better it is. But I felt like I needed to convince myself that this is a true statement, [so] I've decided to fly up there and make sure it's true." And that's what happened: Once aboard *Mir*, courtesy of NASA, he decided that NASA's desire to take the station out of orbit as soon as possible was unjustified.

As the only "space rookie" cosmonaut to fly on a shuttle during these visits to *Mir*, Salizhan Sharipov was an interesting exception to many of the typical Russian rules. At 33, he was one of the most junior pilot-cosmonauts left in the Russian program, and he was pretty close to last in line for a space mission. Even after seven years of cosmonaut training, he wasn't scheduled to make his first flight into space for at least three years, maybe as many as five.

So after completing routine splashdown survival training in the Black Sea in the summer of 1997, Sharipov had scheduled a six-week vacation with his family. For the first time in years, he was going home to his native village in mountainous Central Asia. Although he was a Russian citizen, he was ethnically Uzbek, a Turkic nationality in Central Asia (he was *rossiyskiy*, referring to the Russian state, but not *russkiy*, referring to the Russian ethnic majority). To complicate things even further, when the borders were drawn up, his village wound up inside Kirghizia, a nation dominated by a somewhat different Turkic culture.

But shortly before his departure, he was called into the office of General Yuri Glazkov, the short, chubby space veteran who directed the day-to-day affairs of the Russian cosmonauts at their training center at Star City, northeast of Moscow. Glazkov had a startling announcement.

"We're canceling your vacation," Glazkov told the space rookie. "We want you to go to Houston for a shuttle flight in six months." He explained to the astonished but overjoyed cosmonaut that NASA had invited the Russian Space Agency to send cosmonauts up on the last two shuttle missions to *Mir*. NASA believed that such representation enhanced the value of the mission, provided the safety feature of having a native speaker aboard in case of difficulties, and was an investment in the future for better personal relations between American and Russian space travelers. But for the startled Sharipov, it was a "very unexpected" shortcut into outer space.

After being approved by a special commission and beginning intensive language training, Sharipov packed up for Houston in August (his wife was able to join him in December, but they left their nine-year-old daughter and six-year-old son in Star City to continue their schooling). Aside from the drenching heat and humidity, Sharipov remembers most the "extreme difficulty" of mastering enough English to function with his crewmates. But he made remarkable progress in both his technical training and his language skills. When he met with the news media in January 1998, he impressed everyone with the smooth ease of his conversational English.

When we first met, I tried out the short list of phrases in Kazakhi, a language very similar to his native Uzbek, that I had learned for my visits to Baykonur. His eyes widened with pleasure, and as we worked through each set of carefully preplanned phrases, he jumped to the conclusion that I was much more proficient than I really was. I gently explained the truth, and we chatted on in English and Russian.

Sharipov blasted off in January 1998 aboard *STS-89* with the title of mission specialist. This designation is normally given to astronauts who have completed a full year of generic training and another year of crew-specific training, so there was some question about whether the terminology was an honorary title that was granted for diplomatic purposes.

"We use the term because he is being trained to perform shuttle-related functions," explained Steve Hawley, an astronaut then serving as NASA's deputy director of flight crew operations. "We also give significant credit for cosmonaut training he's already received in Russia." Although the specific hardware on the shuttle is different from that on the *Soyuz* and *Mir*, most of the principles of space systems operations are common to both countries' hardware.

When he returned from the mission, Sharipov had his interrupted personal plans to complete. "My brother keeps bees in the mountains," he had told me before launch. "I love to go hiking up there. It is very wild." The following summer, when the snows had melted and the honey was fresh, Sharipov and his young family made their long-delayed visit to their homeland. He saw it with new eyes, he told local journalists, as the view of it from orbit had been stunning.

Sharipov had almost lost that vacation, too, as the Russian space organizations battled over who would be aboard the first shuttle mission to *ISS*, scheduled for the following December [1998]. Sharipov was the favorite of the Star City training team, but RSC-Energia wanted to send the veteran shuttle cosmonaut Kondakova. Months of useless standoff followed through the spring and summer of 1998, until finally it was decided that Sergey Krikalyov, already a member of the *Expedition-1* permanent crew, would be the Russian representative. In the minds of many Russians, sending Kondakova on her second flight only months after her husband's would make the program too much of a "family affair."

Originally, it was also widely believed that the shuttle commander of the last visit to *Mir* would be Frank Culbertson, Ryumin's American counterpart and the astronaut who was serving as NASA's director for shuttle-*Mir*. Culbertson had never made a secret of his desperate desire for another space flight, and this longing fully guaranteed his loyalty to the NASA officials who selected the flight crews (essentially, George Abbey). He often stated that he and Ryumin had joked about making a space mission together. But the crises of 1997 had taken up the time that should have been spent on Culbertson's shuttle flight–related training. He reportedly took himself out of consideration because of his recognition of the importance of his current assignment. To replace him, after several other pilots turned the mission down, he finally settled on Charlie Precourt, who would be returning to *Mir* only a year after he left on his previous shuttle docking.

In the end, *Mir* had been saved through impressive feats of human endurance and ingenuity. Program veterans were taught lessons that they would never forget. And that—not anything that was actually achieved by the men and women who served aboard it—is probably *Mir*'s greatest accomplishment. Within a few years, it was to look even more impressive. NASA officials would go on to make many of the same mistakes, and relearn many of the same lessons, aboard the *International Space Station*. Among them would be Culbertson, who finally got his desired next space mission in late 2001 as commander of the third expedition to the new station.

10
Lessons Learned
and Unlearned

"Experience keeps a dear school, but fools will learn in no other."

Benjamin Franklin

"What one has not experienced, one will never understand in print."

Isadora Duncan, *My Life*

"On a lark," in his own words, shuttle-*Mir* chief scientist John Uri decided to compare his experiences with those of the NASA researchers on the *Skylab* space station a quarter of a century earlier. This was four years after the beginning of scientific activity aboard shuttle-*Mir*, and a year after such activity had ended. In an insightful memo from July 1999, Uri described going to the data center at the Johnson Space Center and finding a document called "*Skylab* Lessons Learned as Applicable to a Large Space Station." It was a Ph.D. dissertation by William Schneider, who had been *Skylab*'s program manager.

He immediately realized one thing: "I should have read this prior to starting working on *Mir*." To his surprise, he found that "the *Skylab* guys went through a lot of the same things we did on *Mir*." After some major hardware failures during the launch of the station, "the replanning never stopped."

This included "launch accelerations and decays, planning for a rescue mission, addition of new experiments (one included two [space walks]) in the middle of the program, near-realtime lengthening of a mission by a month, loss of a science airlock, replanning activities based on what was learned on previous missions, etc. They also had coolant leaks, attitude control problems, thruster leaks, crew overload, etc."

"Despite all that," Uri continued, "the program was saved from the brink of disaster to become extremely successful."

Even so, Uri concluded, he wasn't sure whether knowing all that *before* the *Mir* program began would have been of any use. "Without having experienced the [*Mir*] program and its challenges first hand," he concluded candidly, "I probably would have missed the point of many of the lessons."

Time had shown that the experience gained from *Skylab* had been useless for shuttle-*Mir*. So how was NASA able to claim that the shuttle-*Mir* (or "Phase 1") experience had been so valuable for *ISS* ("Phase 2")? And did NASA ever learn what the critics had maintained all along: that most of what shuttle-*Mir* had taught would have been known in advance, if only the agency had been properly prepared?

Not surprisingly, the astronauts who were aboard *Mir* were convinced that their missions were crucial to the future and that their experience was essential to the *ISS* program. "I can't imagine proceeding into the *Space Station* assembly without knowing what we've found out on these flights," NASA *Mir* veteran Mike Foale told a news conference in late 1997.

"*Mir* is like having a crystal ball to look at the 'would have been' future of the *ISS*," NASA *Mir* veteran David Wolf told me. "By seeing exactly what systems and precisely which components within them fail, we can specifically target design improvements and prevent similar failures on *ISS*."

Wolf continued: "We can also learn the actual system lifespans and reduce unnecessary maintenance work. In each case this translates to enhanced productivity. In this way the 'failures' on *Mir* have a beneficial feature beyond purely the experience of doing the repair work. We are essentially obtaining the benefit of over a decade of flight testing."

"The prime reason for doing Phase 1 is to learn how to do long-duration space operations," NASA shuttle-*Mir* program director Frank

Culbertson told me during an interview for *Spectrum* magazine in 1998. "I believe we have proven that the things we are learning have direct applicability to operating in space, to operating on a station, to learning how to operate with an international partner."

Valeriy Ryumin, the former cosmonaut in charge of the Russian side of the shuttle-*Mir* program, agreed. "*Mir* is an excellent laboratory for Alpha [Phase 2], particularly now, when the systems have exceeded their usual service lives," he told NASA in a September 1997 briefing. "Nowhere else [c]ould we ... experiment in such real-life circumstances. The experience we will gain together is invaluable."

Others were not so sure. NASA's pioneering spacecraft designer Max Faget, who holds the patents on spacecraft from *Mercury* to the space shuttle, told me that he wasn't surprised by any of the publicly stated "lessons" of the program. "We knew enough not to run cables through pressure hatches back when I was in a submarine in World War II," he added.

Apollo astronaut Walter Cunningham, who helped to develop *Skylab*, NASA's first space station, told me: "I believe we have learned very little that is not self-evident or could not have been found out immediately" once *ISS* began. He granted that there might have been a few lessons in logistics management, but he considered them to be "precious little return for the financial and human resources the United States has invested."

NASA clearly learned lessons in several areas. The agency discovered a lot about dealing with international partners in general and the Russians in particular. It also learned a lot about planning and conducting long-term human space operations, whether or not the Russians were involved. And its familiarity with the operational features of dealing with large space structures—docking, attitude control, thermal control, and so on—was altogether lacking before the shuttle-*Mir* missions.

Some of the stories about the shuttle-*Mir* learning process are certainly good news for the future of the *ISS*. Many of the publicized claims are accurate, and the lessons are valuable.

In 1997, *Mir*'s leaking coolant loop was an excellent example of how on-board experience uncovered unsuspected design problems in the *Service Module* that was being built by the Russians for *ISS*. *Mir*'s thermal control system uses piping that is built into the structure of the station, often along inaccessible portions of the inner hull. After 10 years in space, the lines began leaking, often spraying large amounts of noxious ethylene glycol into the cabin atmosphere. In addition, without the active cooling from these lines, critical equipment that was hooked to inoperable coolant loops could not be used for more than a few minutes without overheating.

"The leaks were primarily the result of dissimilar metal corrosion that required the right combination of time on orbit and increased condensation to manifest itself," Frank Culbertson told Congress in 1997. "Design changes were developed for the Russian modules to be flown as part of *ISS*, to prevent the problem from occurring there."

After recognizing the problem, the Russians had to tear out the coolant loops in their partly constructed *Service Module* and replace them with what they hoped were sturdier lines. It was an excellent case study on the educational value of the *Mir* breakdowns. However, having a U.S. presence on board *Mir* had not been necessary in order to learn the lesson, since the Russian cosmonauts who were there would have noticed the problem on their own. They didn't have to tell NASA about it. Engineers back on Earth could have made the necessary changes to their own design without any American knowledge or advice.

Other design improvements made to the *ISS* as result of the U.S. experience on *Mir* involve the force exerted on equipment by astronauts' movements inside the station. Before *Mir*, designers were using documented values from an old reference book for the landing and push-off forces. The values were based on *Skylab* experiments from 1973–1974 in which the astronauts had been deliberately shaking the whole station as hard as possible.

But careful measurements made on *Mir* in 1996–1997 by U.S. astronauts Shannon Lucid and Jerry Linenger showed that the structural strength requirements designed into the *ISS* were much too high. The finding was confirmed by a few tests on one shuttle flight and by ground tests on the air-bearing (frictionless) table in Houston. The reduced load requirement saves significantly on the weight of most of the internal secondary structures, such as racks, panels, and handholds. If this overkill hadn't been demonstrated on *Mir*, most of the hardware on *ISS* would already have been heavily overbuilt by the time such loads were measured.

The cultural exposure also enlightened both U.S. and Russian nationals, although one cynic noted that it was "nothing that other Americans hadn't learned long ago." NASA's lists of lessons learned contain dozens of cultural observations, although the utility of these observations may be questionable. "Russian engineers within RSC Energiya seem to be genuinely unconcerned about the U.S./Russian contract between NASA and RSA for the NASA/*Mir* program," noted one August 1997 report. "Based upon past experience, the Russian engineers appear to get no benefits from jobs well-done and no penalties seem to be incurred from jobs poorly done." So what else was new?

Among the discoveries about how the Russians do business was the observation that there is no monolithic "Russian space program." Instead, there was a conglomeration of uncoordinated and often antagonistic authorities. "The cosmonaut training center and Mission Control—they hate each other," one NASA controller told me privately in 1998. "You'd think they were in two different space programs." The Russians' wretched pay encourages an ongoing hemorrhage of experienced workers from programs, and this controller added that "the unpublished feeling is that this problem will only get worse."

Surprisingly, officials at NASA headquarters always seemed to be caught unawares by the repeated delays in the deliveries of promised Russian equipment and services. Yet one internal report advised that "generally the accuracy of a schedule can be [estimated] based on time before the event" (or as a countdown to liftoff) with adjustments for predictable delays. According to the report, "L minus 1 year" usually means it may happen in about 21 months; "L minus 1 month" usually means it will happen in two or three months.

Another genuine lesson deals with crew training techniques for long and short space missions. In a special 1998 White Paper on the benefits of the shuttle-*Mir* program, NASA stated that "the U.S. approach to crew training and to scheduling of crew members aboard the *Space Station* is being modified thanks to experience aboard the *Mir* . . . [which has] shown that training for *Station* crew members should emphasize generic skills rather than specific experiment-related skills. This will allow the astronauts to better deal with a wide range of potential circumstances," it concluded.

Culbertson elaborated, "The Russians train for systems skills rather than for flight-specific tasks. This is a natural approach for long-duration missions.... The U.S. crews and instructors are becoming more accustomed to this strategy." Added Wilbur Trafton, a NASA deputy administrator who resigned late in 1997, "We learned quickly that when you are operating a space station, you need to do skill-oriented training, and give the researchers their objectives, and let them go conduct research." NASA administrator Dan Goldin has often cited this discovery as a big payoff of the program.

Outside observers retort that the discovery could have been made without U.S. experience on board *Mir*. There would have been opportunities to learn the same thing in preflight simulations (such as the 90-day isolation runs to test life support systems in 1997), through interviews with *Skylab* and *Mir* veterans, and from actual experiences aboard the *ISS*.

Further, the much-vaunted transition to Russian-style training was soon reversed, at least insofar as space-walk training is concerned. "There were big changes in EVA planning" in February 1998, one NASA Mission Control expert told me. Skill-oriented training was out, and task-oriented training was back in.

The change of mind followed the more thorough planning of the EVAs required to assemble the multimodular *ISS*. In the early 1990s, the plans for space station *Freedom* had called for all assembly outside the spacecraft to be performed by the visiting shuttle crews rather than by the station crew in residence. But the tasks grew so complex and required so many hours to complete that they could not be finished during the week-long period when the shuttle was docked. So they fell to the station's permanent crew.

This shift in responsibility went along with the advertised philosophy of stressing skill-based rather than task-based training. As a result, space station crew members were trained for complex tasks many months before they were required to perform them. Although this worried some NASA veteran planners, it seemed to be the way the Russians were doing it.

In the end, however, it did not work. The strict requirements for extensive assembly training and the long lead time between training a space station crew on Earth and that crew's actual performance of the operations months later in orbit overburdened the schedules of the astronauts and training facilities alike. So by 1998, the assembly tasks had been reassigned to visiting shuttle crews who were fresh from much more specialized training.

Former astronaut Kevin Chilton, the *ISS* deputy director for operations from 1996 to 1998, told me, "We've insisted that the time-critical assembly EVAs be done with shuttle crews" who have been task-trained.

Using astronauts with only generic skill training was simply not feasible. "We saw how skill-based training failed during the contingency EVAs late last year," a Mission Control EVA expert told me in 1998. "The cosmonauts only had to install handrails and braces on the damaged *Spektr* module, but it was much more difficult and time-consuming than had been expected."

In this case, the latest operational lesson learned was to disregard what people had previously thought was a classic lesson learned about the benefits of Russian-style training techniques.

NASA has circulated other arguably dubious claims about the *Mir* experience. During the intense public debate in September 1997 over continuing to send Americans to the troubled *Mir*, the space agency issued another White Paper on the "Benefits of the Shuttle-*Mir* Program." Repeating official assertions about the program's goal of gaining long-duration tour experience, the three-page report stated: "For less than two percent of the total cost of the *Space Station* program, NASA is gaining knowledge and experience through shuttle-*Mir* that could not be achieved any other way."

But the claims often fail to withstand scrutiny. The claim that the experience in rendezvous, docking, and spacewalking was valuable is undermined by the observation that plans for these operations were based on ground analysis and that in flight, the operations worked precisely as originally planned. If NASA had gone directly to the *ISS*, the planning for such operations would have been the same as that for shuttle-*Mir*, and the operations most likely would have worked just as well. There was no need to do them first on *Mir*.

All of the other examples of lessons learned listed in the NASA White Paper involve the refinement of procedures that might well have gone just as smoothly on *ISS* missions as they did on shuttle-*Mir* missions. These procedures encompass target lighting, sensor development, equipment verification, techniques refinement, and the like. Some of the claims had nothing to do with *Mir* at all, but were associated with independent experiments performed on shuttle flights during *Mir* docking missions.

Still, NASA officials kept coming up with new lists of shuttle-*Mir* benefits. Randy Brinkley, then the NASA space station program manager, defended the worth of shuttle-*Mir* experience during a speech at the 48th International Astronautical Congress in Turin, Italy, in October 1997. He described procedural changes made by NASA on the basis of the *Mir* experience. Examples were procedures for operating joined modules, for controlling a large structure using space shuttle jets, and for exchanging data between U.S. and Russian control centers.

NASA scientists proudly pointed to discoveries about the shift of radiation zones in space or to photographs of Russia's diminishing Aral Sea as major scientific benefits of the program. But automated satellites monitor space radiation much more reliably than any human piloted vehicle, and automated satellites can take much better photographs of Earth's surface than any astronaut. So these claims appear desperate, even pathetic.

Other *Mir* veterans made recommendations for the *ISS* in light of their own experience with *Mir*'s shortcomings. Shannon Lucid, who spent six months on *Mir* in 1996, said that the *ISS* needed a much better inventory control and stowage system. "One of the biggest problems with doing work on *Mir* is stowage and being able to find things. Once you found the things needed to do your experiments, you were 90 percent finished." But why hadn't the Russians, or anyone who interviewed the cosmonauts or read their published memoirs, discovered this years ago? German *Mir* veteran Ulf Merbold had told NASA about the problem in 1994, but apparently nobody was listening. And as the experience of the first crews aboard *ISS* in 2000–2001 showed, NASA and the Russians made exactly these same mistakes all over again.

French *Mir* veteran Michel Tognini specified other problems with *Mir* that would have to be fixed for *ISS*. There were too few body restraints, too few opportunities for communication with Earth, and no portable computers. But *ISS* plans already included portable computers and near-full-time communications links. *Mir* merely confirmed that the original plans for *ISS* were correct.

Furthermore, merely listing *Mir*'s inadequacies has never guaranteed that the desired improvements will show up on *ISS*. To illustrate: space flight veterans reported that trash management on space vehicles was an unanticipated daily challenge and would require special procedures on *ISS*. On space shuttle flights, the piles of scrap paper, food wrappings, soiled clothing, and broken equipment took up far more room than the tightly packed original packages, and sacks of trash were stuffed into unused corners and tunnels. On board *Mir*, the problem had been even worse.

The original plans for space station *Freedom* called for a trash compactor, but it was deleted from the *ISS* design. The Russians planned to use one of their modules as a flying dumpster, but U.S. measures for handling the problem on the *ISS* were going to be totally inadequate, according to astronauts at work on this problem in the late 1990s. Artistic visions of astronauts soaring through neat, spacious modules would, according to *Mir* veterans, become nightmarish skirmishes with swarms of trash bags, old documents, and inoperable electronics. That's because adequate interior trash and equipment storage volume was not factored into the baseline *ISS* design.

The inverse of the trash problem is the blank paper issue. One of Goldin's crusades is for paperless offices, even paperless control centers. Worthy though this goal may be on Earth, experience on *Mir* shows that it is wrong in space. "On *Mir*, the No. 1 commodity was paper," a space station manager told me privately. "They were writing on walls, on food wrappers, on clothing. They were begging for more paper in upcoming supply flights." Whether the *ISS* crews will be allocated enough paper and enough storage space for used paper remains an open question, but initial problems on the ISS *Expedition-1* in late 2000 gave no sign that the lesson had been learned. On one *ISS* assembly mission in 2000, the crew was even reduced to writing checklists on their hands and arms.

In March 1998, NASA published its apologia, "The Phase 1 Program—The United States Prepares for the *International Space Station*," which listed more benefits: "Phase I space walk experience has highlighted the need for external station viewing capability," it said. "NASA is working to develop robotic fly-by cameras to assist in *ISS* EVA operations and station inspection." This project was soon canceled, however, in order to save money.

"Monitors on the outside of the Russian station found that *Mir*'s surface is being contaminated by residue from its own attitude control propellant," the report also noted. "In an effort to avoid the same pitfall, *ISS* propellant venting procedures have been changed." But that's entirely a Russian hardware issue, because they're the only ones with thrusters on *Mir* and *ISS*. NASA participation in that lesson was marginal at best.

Sometimes the benefits NASA claimed in public were ludicrously trivial. Late in 1997, Mike Foale was answering questions from kids on the Internet. Will Stephens, age 11, of Evanston, Illinois, asked him what he would tell the people who were planning the new international space station. Foale replied that he was amazed that the bungee cords that the Russians used to stow equipment were so useful, since NASA had always used Velcro to make things stick together. "I believe that the combination of Velcro on equipment and bungees in general on the walls is a very good way for us to keep control of all the different things and pieces that we have to work with when we're in space. And I think that's the most important thing I'm going to tell anybody about the *International Space Station*." Maybe Foale was just giving an example he thought an 11-year-old could understand, but it's still a trivial payoff in view of the expense of the program.

Surely there were more serious and substantial lessons to be learned. NASA must have genuinely thought so, as it assigned a team of engineers to capture them. In one such effort, Jeff Cardenas, a lead operations specialist with Phase 1 (shuttle-*Mir*) in 1997–1998, was assigned to ensure that documented lessons from shuttle-*Mir* were reaching the right audience on the *ISS* side. As he told me in connection with an article that I was writing for *Spectrum* magazine, he had increasingly been hearing the plaintive refrain, "Gee, we wish we knew that two years ago."

"Pockets of guys over there recognize that things have gotten out of hand and that they need some help," he continued. "But most Phase 2 (*ISS*) folks are unconvinced that they have anything to learn from Phase 1 (shuttle-*Mir*)."

One barrier was the way that the lessons were written. "It was highly varied," Cardenas told me. "Often it simply wasn't comprehensible to Phase 2." On the Phase 2 side, deputy director for operation Kevin

Chilton agreed. "The key to success is how a good lessons-learned is written," he told me.

Raw memos or the minutes of *Mir* debriefings are not enough, according to Cardenas's coworker David Lengyel. "These are written in *Mir*-specific terms and on the surface are not salient to Phase 2," he told me. Phase 1 engineers assumed that most Phase 2 workers ignored most of the reports the Phase 1 engineers had filed on the lessons they had learned in their program.

Lengyel, who came to NASA in 1995 from McDonnell-Douglas, St. Louis, and after Phase 1 headed a task force on the transfer of lessons learned, expected work like his to improve matters. A lot of people have a mindset that has to be overcome, he explained. "The Phase 1 people see themselves as stepping stones to Phase 2, but the Phase 2 people in general don't see themselves as heirs to Phase 1."

Again, when I spoke with him in 1998, Chilton did not argue with this assessment. "We have two groups each going 90 million miles an hour who hardly have time to pause and speak to each other," he told me. "The human tendency on the Phase 2 side was to say that the Phase 1 stuff doesn't apply, since we're not flying *Mir*. And the human tendency on the Phase 1 side is for guys to run into that wall a few times and then not want to do it again."

At any rate, in 1997, Chilton got Lengyel to organize several forums of Phase 1 and Phase 2 engineers. One of them assessed *Mir* as an analog of the *ISS Service Module*. Lengyel told *Spectrum*: "We looked at in-flight maintenance, at their spares philosophy [what to do when a part exceeds its rated service life, and which spare parts would be stored on board for immediate replacement and which would be launched on later flights only on demand]." The Russians tend to operate systems until they break, and NASA has moved toward that philosophy instead of following the scheduled change-out strategy that was originally planned. According to Chilton, Phase 2's lack of respect for Phase 1's "lessons" stemmed in part from the fact that some of them had already been learning the *Mir* systems as they related to the Russian *Service Module*. In some cases, the Phase 2 engineers knew the Russian systems better than the shuttle-*Mir* guys, who had concentrated on the experiment operations, at least before the 1997 accidents.

And sometimes the Phase 2 folks were justified in their skepticism. "Surprisingly, some of the Phase 1 types did not actually learn all the lessons we would assume," an engineer who had worked in both programs told me. "Certainly they learned a lot of things, but the political landscape of *ISS* is much different. Same players, different game." As an example, he cited the interplay between the different Russian organizations, such as Energia and Khrunichev. This was crucial for *ISS*, but Phase 1 never experienced it because they dealt exclusively with Energia, and they had the "big stick" of NASA money that they could use to get their way. For the *ISS* project, Russian cooperation had to be cajoled.

Lengyel concentrated on measurable lessons. He noted that the Russians are orbital pack rats, stowing away all sorts of broken equipment just in case any still-serviceable components are ever needed. "The Russians tend to cannibalize, they keep parts on orbit," he explained. These useful insights dealt less with hardware design than with operational philosophies. "Ninety percent of these lessons are process oriented," said Lengyel. Chilton used the term *methodology* to imply the same thing. "Some are technical, or cultural, or a combination. And it's not just U.S.-Russian cultural gaps, it's those within NASA, or between NASA and contractors."

As the teams and their documentation matured, more opportunities to effectively share experiences arose. On the *ISS* side, a detailed 10-volume document called the Station Program Integration Plan (SPIP) went through numerous iterations until, by late 1997, it began to assume a usable form. The documents defined the processes and subprograms within *ISS*, as well as the relationships among the organizational structures, their responsibilities, and their resources. "We saw an opportunity to apply our lessons in a very systematic and analytical way," Lengyel explained. Now an expert could take a lesson from Phase 1 and determine exactly who in the Phase 2 world was responsible for implementing it. "It defined the targets for our lessons-learned," Cardenas added.

These new efforts, as well as the wisdom that comes with experience, have resulted in the transfer of more and more operational lessons. Noted Cardenas: "Their first reaction was 'we aren't going to have that problem,' but now their second reaction is 'we never thought of that.'" From the Phase 2 side, Chilton corroborated that impression: "We weren't smart enough to ask ourselves the right questions until we saw someone else trying to solve something we never thought would happen."

As an example of Phase 2's newfound willingness to learn from shuttle-*Mir* experience, Chilton pointed to the issue of planning for the transfer logistics. Aboard the *Mir*, visiting crews unloaded tons of supplies by hand, package by package, and all with proper paperwork. At first, the Phase 2 side claimed that on the *ISS*, the crew would transfer equipment rack by rack, not piece by piece. But more detailed analysis showed that there would be a need for many piecemeal transfers as well, said Chilton.

Beyond these practicalities, Chilton and the other operations specialists stressed the immeasurable benefits of using the Russians' experience and practicing with them on *Mir*. For developing trust, studying is not enough. "You've got to be working problems together" to get to know each other, Chilton said. He admitted that it took time, but he concluded: "We've got that slow start behind us in Phase 1 instead of an extra risk on Phase 2."

It wasn't just the *ISS* program that learned things. The space shuttle program picked up some important new tricks as well. "Phase 1 taught the shuttle program to be far more flexible than it has ever been," a friend told me in 1998. "The program took the long downtime this summer to look at all the lessons learned from the successful scrambles to put large equipment aboard at the last minute (tools, Vozdukh air purifiers, Gyrodynes, computers, even alternate crew members), and has developed a strategy for a 30-day mission planning cycle. What used to take two years by the book now has an actual plan to assemble a mission and put it in orbit in 30 days. Flexibility is the key to success and savings, and I think that the Russians taught us a lot about how to prepare for a continuous presence in space."

Phase 1 wasn't free, of course. If the amortized costs of ground operations and nine shuttle flights are included (although NASA does not keep its books that way), the entire Phase 1 program cost more than $5 billion. NASA had also canceled a series of scheduled shuttle-*Spacelab* missions to clear the way for the *Mir* dockings. Aside from the value of the data that would have been obtained (and must now be delayed for many years until science operations on the *ISS* begin), these missions were expected to test U.S. hardware intended for installation aboard the *ISS*.

"A lot of things could have been learned in the *Spacelab* flights," I was told by a NASA *ISS* manager. "They were going to fly *ISS* racks and a lot of other hardware—cooling systems, air revitalization systems, avionics racks, and so forth." Following their space shakedown missions, these refrigerator-sized racks would be modified as needed and then taken back into space for permanent installation on the *ISS*. If this hardware isn't flight-tested, it must be produced right the first time. The initial flight testing on shuttle missions had been expected to detect first-flight troubles with such racks. Since they were now to be installed aboard the *ISS* without such testing, having the troubles occur then instead of earlier is yet another cost of flying shuttle-*Mir* missions instead of the originally planned rack missions.

If many of the obvious results of shuttle-*Mir* were never implemented, the root cause may just have been the schedules. By the time much of this information got to the people who needed it, the designs for the *ISS* were already complete. "The design of the international space station has been frozen for some time and flight hardware is being built," former *Apollo* astronaut and U.S. Senator Jack Schmitt wrote in late 1997. The designers were really getting smarter, but usually, it was too late.

My friends in the program agreed. " I wouldn't say that we overlooked lessons learned," one e-mailed me. "I think that in most cases we don't have the money in the program to rebuild, where we now know better. At some point, 'better' DOES become the enemy of 'good enough.' Most of the lessons, unfortunately, have come too late to roll into the Phase 2 hardware, but quietly, Phase 3 (advanced *ISS*) is rolling up the lessons learned, because their TIME axis is a little larger. They say that experience is that part of your education that you get just after you need it most."

So for the *next* time, there are probably even more lessons that can be learned from *Mir* (and from *Skylab*). They include principles such as:

- Design evolution, not revolution. Over the history of *Salyut* and *Mir* and its modules, almost every new module either was an improved version of an earlier one or used subsystems from a previous vehicle. This allowed innovations to be introduced gradually and created a generally tolerable level of anomalies. But for all the modules that the United States is building for *ISS*, there is no previous flight history. This is almost a guarantee of an avalanche of in-flight surprises.

A NASA engineer involved with the *FGB*, the Russian-built module that was the first piece of the *ISS* launched into space, agreed that iterating designs generation by generation was advantageous. "The reason we have had success in integrating our U.S. systems inside an *FGB* is that the *FGB* is in its 9th generation, with a high success rate, and because it is designed to be a flexible platform which you can customize as needed," he explained. And using the *FGB* "was a particularly insightful move: it put the location of all of the structural, electronic, and environmental interfaces between the major partners in the most robust, well-debugged, and well-understood platform in the whole program. We'd have been doomed, trying to do this from scratch, or conversely, trying to cram our U.S. systems inside a module that is so completely potted, like the *SM*."

- Be operational at first launch with a working life support system. Getting a crew aboard permanently, and early, to allow rapid bootstrapping (tech talk for self-help, where the space station actively supports its own assembly) of the vehicle. That way the station will not be hostage to component slips downstream that would otherwise cause major delays in operational status. It will also be necessary to expand the station as much as possible, over a period of years, to augment basic capabilities. This principle was violated on *ISS*, and NASA paid the price for it: years of delay, extra shuttle missions to service faltering space hardware, and public embarrassment.

- Keep the great minds focused on the future and don't staff new efforts with "leftover" personnel. As soon as any Soviet space system became operational, the originating agency farmed out any follow-up operations to spin-off organizations, keeping its brightest people to work on the new projects. Korolyov himself started this Russian tradition. Like a space Johnny Appleseed, he sowed space and rocket institutes all across Russia. But it is commonly felt that the opposite happened in the United States. With the exception of a core cadre of true enthusiasts, the people assigned to the *Space Station* program over the last 15 years, from top to bottom, were the ones that the operational organizations didn't need or want.

A top NASA manager shared these concerns with me in 1998. He spoke about what he called "the biggest problem NASA has always had in its Space Station plans," and he recommended that NASA follow the policy "that the Russians and almost all successful government and business organizations around the world follow: Put your best people, most experienced people on big, new projects! Most of the people in NASA space station management and Program Office grunt slots have always been castoffs from their former organizations. Or [Mission Control] people or astronauts—smart people perhaps, but what experience from sitting at a console (even the Flight Director console) or flying the Orbiter qualifies a person to manage a HARDWARE DEVELOPMENT PROGRAM???!!!"

He went on to say, "This is the root cause of all the NASA Space Station problems—the wrong people (i.e., people with inadequate or wrong qualifications) have always been in charge. And this is showing itself now, more and more each day."

An anonymous comment posted on *NASA Watch* in October 1997 made the same point in response to an earlier comment that many "refugees from Seattle [Boeing]" had moved to the Johnson Space Center (JSC) to learn how to do space station work. "I note that there were also many refugees from [the Boeing Space Station office in] Huntsville [Alabama] that moved to JSC—those were the ones that couldn't get jobs to be able to stay in Huntsville!"

My own experience didn't leave me quite so cynical, because I know a lot of the people working on the *ISS*. I know that they joined up because they were dedicated to the project, not because their original organizations didn't want them. But NASA always has let the operational experts run the current astronaut-related programs, while any future programs were scoped out by extra (sometimes unwanted) and very junior workers. Then, when the current program was phased out (*Mercury, Gemini, Apollo, Skylab* were all handled the same way), the senior operations people moved on to the next project

Bundled against the cold of space in the derelict *Salyut-7* space station in 1985, Viktor Savinykh works to restore power in what was the zenith of Soviet cosmonaut success. To everyone's surprise, years of decline and decay lay ahead of the Russian space program. (*Courtesy of USSR Embassy*)

Konstantin Feoktistov, one of Sergey Korolyov's brilliant young lieutenants at the beginning of the Space Race, was a member of the world's first multicosmonaut space mission in 1964. He later headed his own design bureau before being forced into retirement in 1990 after the death of his patron, Valentin Glushko. Feoktistov came to believe that Soviet human space projects were extravagant and useless. (*From the author's collection*)

Russia's "rendezvous ace" Vladimir "Johnny" Dzhanibekov was the USSR's hottest space pilot of the 1980s and by all accounts an exceptionally decent human being. His superb piloting skills made the rescue of the *Salyut-7* space station possible, but within months it was all thrown away. He never flew in space again, and soon after was nearly killed in a severe automobile accident.

Cosmonaut Anatoliy Solovyov was Russia's ace space traveler in the 1990s, setting records for total numbers of *Mir* visits, space walks, and crises conquered. Selected for the first *International Space Station* crew, he turned it down rather then serve under a less-experienced American astronaut. Retiring as a cosmonaut, he campaigned fruitlessly to save the *Mir* space station from destruction. (*Courtesy of NASA*)

Paintings of the massive *Polyus* spacecraft (a prototype Battlestar weapons platform launched in 1987) hang on walls of program managers at the Khrunichev Institute, which later sold a modernized *Polyus* rocket module to NASA as the cornerstone of the *International Space Station*. (*Courtesy of the Khrunichev Institute*)

Pioneers of the Soviet space program, first human in space Yuriy Gagarin (*left*) and founder of the Soviet space program Sergey Korolyov (*right*), in 1961 soon after Gagarin's flight. Korolyov died in 1966 at the age of 59. The Soviet space program stumbled, but picked up steam again under his successors in the late 1970s and early 1980s. Gagarin died in a plane crash in 1968.

ISSN 0233-3619

ЭНЕРГИЯ
ENERGY
ЭКОНОМИКА·ТЕХНИКА·ЭКОЛОГИЯ
6'92

ОТЕЧЕСТВЕННАЯ
КОСМОНАВТИКА
НА ПЕРЕПУТЬЕ
стр. 63

As the Russian space program approached bankruptcy early in the 1990s, caustic humor often portrayed cosmonauts as decrepit beggars. (*Courtesy of Energia Magazine*)

The Kosmos Pavilion at a Moscow park was once filled with space hardware and optimistic crowds. Now the giant portrait of Yuriy Gagarin looks over a lost space capsule and rows of automobiles and sailboats for sale.

As part of endless money-making activities, cosmonauts aboard the *Mir* space station accepted contracts to advertise Western products from soft drinks to milk to pizza. (*Courtesy of NASA*)

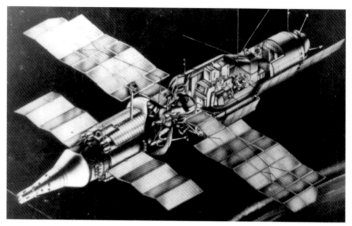

The USSR's *Salyut-3* space station with crew (1974) carried a modified aircraft cannon in its nose (*at right*) to ward off feared American space attacks. In the original design, a cargo ship also carrying crew aboard (*at left*) and with its own landing capsule would periodically dock to the military space station. This vehicle was replaced by the *Soyuz* design, but the supply section was later used as the *FGB* module, the first piece of the *ISS* launched in 1998. (*Courtesy of the Khrunichev Institute*)

Nudelman air-to-air cannon designed for Soviet jet fighters was adapted for use in space and installed aboard the *Salyut-3* space station. It was test fired by remote control from Earth. (*From author's collection*)

In a survival kit inside each *Soyuz* spacecraft is a three-barrel survival gun, for firing rifle bullets, shotgun shells, or flares (ammunition shown in belt packs). Its folding shoulder stock also serves as a machete. The only weapons aboard space stations, these survival guns have never been used.

Cosmic "rodeo" in 1994 with *Mir* and four visiting space vehicles. An expended *Progress* supply drone (*at left*) is backing away from station's forward docking port to make room for an approaching *Soyuz* carrying crew (photo was taken from the *Soyez*). Another *Progress* drone, still being unloaded, is at the rear docking port (*far right*), and the *Soyuz* which brought the crew currently on *Mir* is docked to the *Kristall* module off to one side (*at bottom*). (*Courtesy of Vladimir Semyachkin*)

Blast-off of a Russian *Soyuz* rocket carrying a three-seat spacecraft of the same name. Every six months a fresh *Soyuz* lifeboat must be brought up to the *ISS* and the expired one brought back. Sometimes the *Soyuz* crew also stays aboard the station for months, but usually the crew serves only to provide short-term taxi service, with two professionals bringing along a third guest, either a visiting scientist or a paying tourist.

For docking their own shuttle to the *Mir*, Soviet engineers improved on the androgynous (neuter) docking mechanism used in *Apollo-Soyuz* in 1975. The Soviet shuttle never used it, but the design was adopted by U.S. shuttles for dockings to *Mir* and the *ISS*. (*Courtesy of NASA*)

NASA's *Atlantis* space shuttle pulls away from *Mir* in July 1995, completing the first U.S./Russian docking in 20 years. The module pointed toward the shuttle, *Kristall*, had been rotated from a position pointing directly out of the photograph. The photograph was taken by cosmonauts in the *Soyuz* spacecraft, which had undocked from the far right end of *Mir* specifically to get this view. At this point the *Mir*'s computers crashed, and it began drifting out of alignment, forcing the cosmonauts to perform an emergency redocking. (*Courtesy of NASA*)

The first Russian-American space handshake in 20 years, the shuttle-*Mir* docking on June 29, 1995, began an intense period of cooperation between former Space Race rivals. *Mir* commander Vladimir Dezhurov (*top*) later served aboard the *ISS*; *Atlantis* (*STS-71*) commander Robert "Hoot" Gibson retired to become an airline pilot.

A typical Russian solid fuel oxygen generator (SFOG) canister like the one that burst into flames aboard *Mir* in February 1997, threatening the lives of all six men aboard. Previous crews had also experienced fires with similar gear but had kept silent about them. (*Courtesy of NASA*)

To relieve tensions in the hours after the fire that nearly killed them, doctors ordered the crew to partake of some medicinal brandy kept on board for such purposes. Linenger, who took the photo, abstained. Program director Frank Culbertson ordered this photograph withheld from publication, but it was released following a Freedom of Information Act process. (*Courtesy of NASA*)

The collision with the off-course supply ship in June 1997 severely damaged one of the solar power arrays on the *Spektr* module attached to *Mir*. It also bent the mast and cracked the module's hull, letting all its air out. Although the *Spektr* laboratory was never used again, cosmonauts were able to repair most of the other damage to *Mir*'s power system. (*Courtesy of NASA*)

Congressional hearings in September 1997 sought to determine if it was safe to send more Americans to *Mir* following the string of unanticipated accidents. From left, Frank Culbertson, astronaut and director of the shuttle-*Mir* program; Roberta Gross, NASA's Inspector General; the author (not standing on a box—is truly 205 cm tall). (*Courtesy of Florida Today*)

At the crossroads of the *Mir* complex, passageways lead forward and aft as well as outward to four research modules mounted like radiating spokes. When one of these radial modules was punctured during a docking accident in 1997, the crew had to quickly remove all the lines and hoses leading into it, and block it with a safety hatch (*lower left foreground*).

Cosmonaut floats in the main hall of the *Mir* space station in the late 1990s. A decade's accumulation of gear and supplies had taken up all the space behind the panels and much of the space in the crew's living areas. Stashing equipment and then finding it when needed was a daily challenge, and attempts to develop more efficient techniques for *ISS* met with only limited success. (*Courtesy of NASA*)

From the top hatch of the *Soyuz*, the commander's seat is shown, facing the control console. Additional seats are to the left and right, and behind the three seats is the main parachute compartment. (*Courtesy of NASA*)

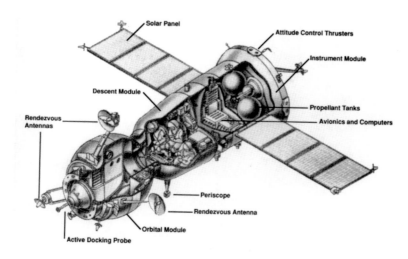

Cutaway drawing of seven-ton *Soyuz* spacecraft shows center module where crew sits for launch and landing, plus orbital module in front, with more living space. Since 1966, more than 100 vehicles of this type have flown in space, and with minor modifications this will be Russia's only human piloted space vehicle for the foreseeable future.

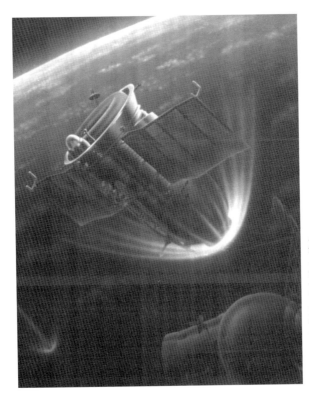

Artist's concept of flame-shrouded *Soyuz-5* spacecraft entering the atmosphere backward in January 1969. Without its heat shield pointed in the right direction, the spacecraft would have burned up within moments. (*Courtesy of Newton Magazine, Tokyo*)

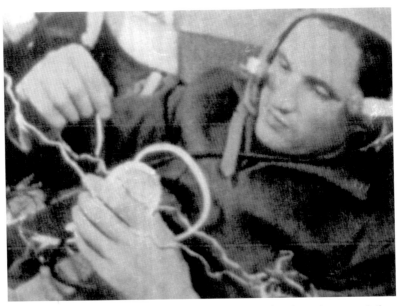

Cosmonaut Boris Volynov was alone aboard the burning spacecraft. Earlier in the mission, before spacewalking to another *Soyuz*, one of his fellow cosmonauts had taken this photo of Volynov at his control post. (*From author's collection*)

Soyuz spacecraft splits into three sections after heading back to Earth. Center section is the shielded command module with the crew. Spherical orbital module in front carries living space and docking equipment. Cylindrical aft section carries rocket engines, life support supplies, and electrical power system. In 1988, a computer error nearly caused a *Soyuz* to separate while the vehicle was still in orbit, which would have led to the death of the crew. When NASA considered using a *Soyuz* as a bail-out capsule for the *ISS*, the Russians omitted mention of this near-catastrophe. (*Courtesy of Space Art International and Andrei Sokolov*)

Soyuz crew cabin on the ground in Kazakhstan, after parachute landing cushioned by soft-landing rockets

Russia's most powerful rocket, the *Proton*, carries the 20-ton *FGB* into orbit on November 20, 1998, to become the cornerstone of the *International Space Station*. The United States paid for its construction, and got a 4-inch flag painted on the side of the module (and a larger one painted on the rocket, which burned up 10 minutes after launch). Russia paid for the launch and kept the module's radio command codes to itself. (*Courtesy of NASA*)

The first two sections of the *ISS* were hooked up in December 1998. The solar-paneled *FGB* provided stabilization and communications, and the inert U.S. *Node-1* just went around in circles waiting for the next sections. Three shuttle flights docked to this rump complex to perform repairs for various breakdowns and to stockpile supplies, but the first permanent crew did not arrive until November 2, 2000. (*Courtesy of NASA*)

Behind the cosmonaut center at Star City, a small village of million-dollar mansions is springing up for senior cosmonauts whose official salaries were only one-tenth of the amount needed to afford them. Throughout the 1990s, White House officials insisted they did not exist, and NASA officials insisted it was none of our business even if they did exist. (*Courtesy of Byron Harris*)

Television news reporter Byron Harris (*right*) confronts cosmonaut deputy commander Yuri Glazkov (*left*) about how his house was funded, during a visit to NASA in Houston. NASA's response was to clamp down on news media access to the Johnson Space Center.

Space flight simulator in Moscow, where test subjects are isolated for months-long experiments in international psychology. In January 2000 a Canadian woman complained of sexual assault from her Russian commander, but she was not allowed to leave the chambers until she had signed a report repudiating the complaint.

Bill Shepherd, commander of *Expedition-1* aboard the *ISS* in late 2000, used his extensive tool kit to rig a makeshift table when the supply mission carrying the regular table was delayed. (*Courtesy of NASA*)

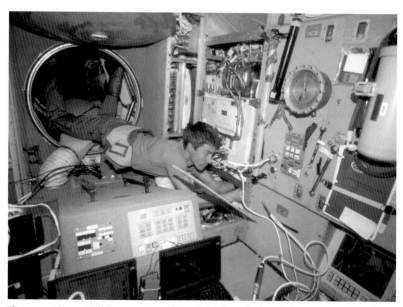

Cosmonaut Sergey Krikalyov, member of *Expedition-1*, works aboard the service module. The first spacefarer actually born during the Space Age (after the launch of *Sputnik* in 1957), Krikalyov's space career is probably barely half over and may include far more ambitious missions over the next 15 to 20 years. (*Courtesy of NASA*)

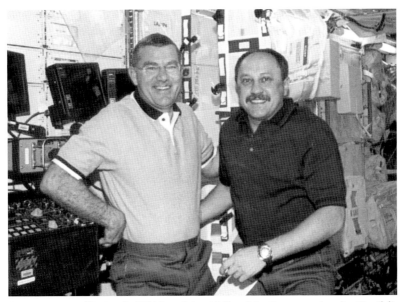

Spacious interior of *Destiny*, the U.S. lab module, with *Expedition-2* crew members Jim Voss (*left*) and Yuri Usachyov (*right*). Behind Voss is the control panel for the Canadian robot arm. Two dozen racks of control computers and experimental apparatus are aligned along walls, ceiling, and floor.

The main windows on the *ISS* face downward toward Earth for photography. This window in the service module shows an approaching space shuttle; a bigger window in the U.S. laboratory shows the same view. The crew has complained about the lack of windows pointing in other directions.

The main structural layout of the *ISS* had taken shape by early 2001. Four cylinders (front to back, the lab module, the node module, the *FGB*, and the service module) were hooked end to end to end to end, some sprouting solar arrays. A towering truss structure extended out the side from one of them, where it sprouted even larger solar arrays and a heat-dissipating radiator panel.

Spacious interior of Italian logistics module *Leonardo* (brought up and down aboard visiting shuttles) gives good impression of the size of the space station rooms when they are not crammed with equipment. Yuriy Gidzenko floats at far end. Italy has agreed to modify one of these modules to serve as a Hab for living space for a larger *ISS* crew.

and took over the leading roles. Since the shuttle has *not* been phased out, this move has not occurred. So when Phase 1 ended, most of the workers were refused transfer to the ISS and either returned to the shuttle or were laid off.

- Design for random attitude tolerance, allowing the station to be in free drift for up to 24 hours or more. This may be needed in emergencies or for routine maintenance. But on the *ISS*, the power and thermal loads for out-of-attitude contingencies have hardly been calculated, much less built into the whole vehicle. Some systems are thought to be damaged by overheating and others by freezing if the station's rigid orientation is lost.

  According to some space station engineers who spoke with me privately, the *ISS* has been verified to be thermally safe only for orientations within 15 degrees of the planned orientation in space. Beyond that, I was told, "nobody knows what could happen." Some components could freeze up and others could over-heat, while the batteries struggle to handle the station's power consumption.

  "The thermal attitude issue is going to be a killer, but we came to that one late, after taking our Freedom hardware from 28 degrees inclination, where it was originally designed, up to 51.6," an insider told me. "Our beta angles [orbital orientation relative to the Sun] then get horrendous. The Russians have been flying there since the start, so we're prepared. We changed plan midstream, and a lot of our hardware picked up robustness requirements that at that point we're hard to meet. If you'll recall, we weren't given any budget to refine our hardware in this exercise. The name of the game was 'slash the budget, or die.' This tends to lead to long range Ops workarounds, rather than sound engineering investments."

- Rely on an inexpensive automated Earth-to-space cargo vehicle for primary logistic support and for reboost, as well as for garbage disposal. The Russian *Progress* system has carried out about 100 space resupply missions, every one successful, at a cost per deliv-ered pound of a fifth to a tenth that of the shuttle. But for the *ISS*, most supplies will be carried on expensive space shuttles, while Russian promises of dozens and dozens of *Progress* missions are fading away.

- The Russians are serious about medical and psychological screening of the crew, and they really do drop people from crews or even from the cosmonaut program for inadequate performance, something that's exceedingly rare in the United States. Their screening tests have been expanded over the years as they have encountered unexpected medical and mental complications on long missions, and their psychological support teams are dedicated to helping crews get through crises, not catching them at human weaknesses. "After each medical problem on board, they invented a new constraining criteria," a French astronaut told me. "After the kidney stones in orbit, they invented the 'floating kidney criteria.' After the toothache in orbit, they try to extract all teeth with root canal before flight. After a prostate infection, they invented the prostate fluid sample draw test."

- Use the metric system; everybody else does.

- Stress in-flight serviceability and repairability of all critical components, an area in which the Russians excel. "Russian cosmonauts can repair anything in orbit provided they have the parts," an astronaut told me. "A good example is the [space] suit, where they can change pumps and valves. I am not sure we changed the U.S. suit to provide the same capability in orbit."

One of the astronauts who made a space walk on *Mir* told me that the way the *ISS* airlock worked was directly opposite to what he had seen work so well on *Mir*. "The [space walk] hatch would optimally be on 'end' of cylindrical airlock, not on wall, to allow better egress/ingress and handling of large gear," he e-mailed me. This stipulation was violated because of "a cost constraint induced by using a modified shuttle airlock, a conscious decision."

The *Mir* experience wasn't a total loss, he insisted. "Note that a surprisingly large number of modifications, learned from phase 1, were actually accommodated, like placing a 'porch' for [space walk] staging just outside the egress hatch." He also had seen new procedures added for handling small item stowage on the walls, for galley/toilet placement, for timeline and communications management, and for other hard-learned lessons from his flight

In a process in which communication is the key ingredient, it's fitting that some important lessons were learned about space-to-Earth communications. "We did learn that uplink is worth the investment," my *ISS* enthusiast friend told me. "George [Abbey] did a good thing by investing in Early Comm [a late add-on radio link inside the U.S. *Node-1*]. It fixed a big hole. If the Russians could afford it, they would put more geo[stationary] satellites up, for longer range coverage. The *Mir* fire taught us that we really don't want to be out of comm very long.

"Conversely," he pointed out, "our operators have started to learn that you don't need a stream of telemetry *every* second to control a space mission: Mir is out of comm for hours at a time, and seems to stay in orbit. We spend a lot of money tracking data because it is there. It'll be interesting to see how much monitoring we actually do after 10 years of steady ops."

He was upbeat about the learning process. "On a program with *this* much inertia, with so much bravado on both sides, it's gratifying that we actually were able to pick up some lessons at all," he concluded.

The Russians have a proverb, "Better to see once than to hear a hundred times." NASA proved it true. In boasting about how much the *Mir* experience taught them, NASA officials unwittingly highlighted the shallowness of their preparation for *Mir*, which led to their astonishment over the 1997 *Mir* crises. When NASA's teams began operating the *ISS* in 2000 to 2001, they discovered that those groups that had lived through *Mir* really performed more smoothly, while those groups that merely read the reports of the first groups blundered and staggered until they, too, caught on. So the biggest lesson learned from shuttle-*Mir* should be that NASA did not need shuttle-*Mir* to learn most of its lessons, if a way could be found to really learn from somebody else's experience. NASA had better develop a better way of learning the lessons that are critical to the future success of the *ISS* and the rest of the U.S. space program, or every generation will have to keep making the same mistakes over again.

11

Soyuz Secrets

"All the history books which contain no lies are
extremely tedious."

Anatole France, *The Crime of*
Sylvestre Bonnard, 1849

"Nothing is so firmly believed as what we least know."
Michel de Montaigne (1533–1592),
Works, chapter 31, "Of Divine Ordinances"

In the delightful film *Young Frankenstein*, Marty Feldman (playing Igor) insists to Gene Wilder (playing Dr. Frankenstein) that his name should be pronounced to sound like "eye-gore." Anyone who knows their Dostoyevsky, however, realizes that real Russians pronounce it "ee-gore." In the final analysis, the people who own the names should be able to determine just how they are pronounced.

So it was a good sign when in the mid-1990s, Americans working within the U.S.-Russian space partnership gradually started to correct their pronunciations of the names of Russian spacecraft. "*Mir*" was a surprisingly tough one; it often came out "murr," like the Christmas gift from the Wise Men ("myrrh"). Individual by individual, I watched and encouraged the transition to "meer" (more precisely, "m'yeer"). It seemed to be hopeless, however, to object to the widespread spelling of it all in capitals (MIR), as if it were an acronym instead of a proper noun, and this practice continues to this day.

But "*Soyuz*" (the word for "Union," as in "Soviet Union") was and still is a problem for American tongues. The *y* is deceptive. In the standard orthography for transliteration from Cyrillic into Latin letters, the *yu* stands for a specific Russian letter, so it can't be split. This makes the syllables so-yuz, not soy-uz, as everyone seems to assume.

There is a lot of leeway in pronouncing each syllable. "So" with a long *o* seems straightforward, but since the spoken stress is on the second syllable, Russians tend to slur the unstressed *o* into a neutral *ah*. "Yuz" has a long *u*, as in "ooze," not a short *u*, as in "buzz.' Also, in speaking, voiced final consonants tend to become unvoiced (the *z* shifts to a sound that is more like an *s*), but this isn't too important.

Try: For "*Soyuz*," say "sah-YOOZ," or if you're even more ambitious, "sah-YOOSS." But please, drop the "soy" sauce, and don't make *Soyuz* rhyme with "coy fuzz."

Why does it matter? Proper pronunciation of other people's names is always a sign of respect, and it will always be appreciated. And the *Soyuz* spaceship also deserves respect, since it's playing a key role for the *International Space Station.*

In the early 1990s, as NASA was drawing up its plans for a space station, there was a problem with leaving a crew aboard the station after the visiting space shuttle returned home: They had no way to return to Earth in case of an emergency.

Although NASA did not have any human piloted spacecraft available (the last *Apollo* capsule had flown in 1975), the Russians did. It was the *Soyuz*. After the collapse of the Soviet Union in 1991, the United States began to explore the possibility of expanded cooperation with the Russian space program. Using a modified *Soyuz* on the American station was a promising idea for a project called ACRV, or Assured Crew Return Vehicle.

Lockheed and Martin Marietta had been developing competing designs for such a vehicle when the Russians entered the picture. There was a veteran *Skylab* astronaut on one of their teams, and he recalled what happened next: "Suddenly in mid-1992, we got word from NASA that we were to go to Moscow to 'help NASA assess the usefulness of *Soyuz* as a crew return vehicle,'" the astronaut (an old friend who requested that his name not be used for fear of NASA retribution) explained to me.

"We went, and assessed that *Soyuz* did *not* meet most of the requirements we were working to. NASA, of course, assessed differently, and made *Soyuz* the baseline," he continued, concluding that the "*Soyuz* still won't hack it." That was because it was too small for many American crew members and too small to carry the medical equipment that would be necessary in the event of a medical emergency. The landing G-forces were too high. The landing "footprint" (the area over which a descending capsule might stray) was much too large for the landing sites that NASA had been studying. The ACRV was supposed to have a three-year on-orbit standby lifetime, but *Soyuz*'s was only six months (the Russians promised to make it a year, but they never did).

The Russians offered to make some modifications, but as they wanted $250 million to do it, NASA dropped the idea. That was in mid-1992. Two years later, with a new president and a new NASA administrator, NASA was ordered to change its mind and decide that the *Soyuz was* adequate.

The basic *Soyuz* spacecraft contains three connected modules. In the center is the *Descent Module*, in which the crew rides during launch and landing. Attached at the back end of this module is the *Equipment Module*, which contains the rocket engines and power supplies for the space vehicle. Attached at the front end is the *Orbital Module*, which contains additional living space, radios, and the docking equipment for linking up with other space vehicles.

The Russians liked the idea of the Americans using *Soyuz* for their space station, since the Americans would be paying cash. But to meet NASA's requirements, the Russians had to turn over a lot of details about the safety and reliability of the *Soyuz*. This went against generations of Soviet-era habits, and against Russian cultural practices, too.

The Americans knew that the *Soyuz* had had two fatal accidents, in 1967 and 1971. But since that time, despite many known malfunctions, it had flown more than 70 times, and each time, the crew had always returned safely to Earth. Russian officials told NASA that this proved the safety and reliability of the *Soyuz*, so please send the money.

Eventually, as Russia became a full-fledged partner in the redesigned *International Space Station*, it agreed to provide a *Soyuz* that would remain docked at the station as an emergency escape vehicle for the crew. Every six months, the *Soyuz* would be replaced with a fresh one.

Officials in both countries asserted that the *Soyuz* was safe, despite the small and infrequent malfunctions that the Americans learned of through official Russian reports. However, the official Russian reports proved frighteningly incomplete; at best, they were misleading. For example, it wasn't until 1996 that the worst "almost fatal" *Soyuz* landing incident became known, and then it was learned only through a Russian newspaper article. The official Russian space agency never mentioned it.

The near disaster had occurred in January 1969. Aboard *Soyuz-5*, cosmonaut Boris Volynov was returning to Earth alone, after his ship had docked in space with a *Soyuz-4* and transferred two of his crew to the other vehicle. The Moon race was in full swing, and the *Soyuz* double flight had tested the space-walking techniques that were to be used by Soviet cosmonauts during their planned Moon landing—hopefully, ahead of the Americans.

The 34-year-old rookie space pilot had completed the course change back to Earth, and the *Soyuz* spaceship then was supposed to jettison its extra modules so that the headlight-shaped *Descent Module* could enter the atmosphere safely. Flying over the South Atlantic, headed northeast, Volynov expected to be on the ground within a half an hour.

But then something went very badly wrong. The spaceship's *Equipment Module* failed to fully separate. The explosive "separation bolts" had fired, but when he looked out the spacecraft window, Volynov was horrified to see the whip antennas from the module still extending past the window. The module was still attached, as Volynov confirmed from the feel of the ship as he tried to turn it manually. "No panic," he whispered to himself once, and then again, according to interviews published decades later.

He radioed his situation to a tracking ship below him, and it passed on the terrifying news to Mission Control in Russia. It only took a few moments for Mission Control to realize that nothing more could be done. The ship's heat shield was at the base of the *Descent Module*, now blocked by the balky *Equipment Module*. Unshielded portions of the vehicle would now be exposed to the 5000 degree Celsius heat of atmospheric entry. This would destroy the entire capsule and its pilot.

Journalist Aleksandr Milkus added a poignant detail to this drama in his account, which was published in *Komsomolskaya Pravda* on April 10, 1998. "In the Mission Control Center one of the officers took off his cap, put three rubles into it and sent it further. Little by little it was filling up with money. This had already happened once, on the day when Vladimir Komarov died" (two years before). Other controllers, knowing themselves to be utterly helpless, buried their faces in their hands.

For Volynov, doing nothing at such a moment was impossible, so he continued to make reports into his voice recorder and to write notes in his flight log. He thought of his thirty-fourth birthday party in his apartment, only a few weeks before. "It's hard to describe my feelings," he said later. "There was no fear but a deep-cutting and very clear desire to live on when there was no chance left."

The spaceship was falling back into the atmosphere, and he began to hear grinding noises as the deceleration stresses built up. The ship was slowly tumbling end over end, exposing its entire surface to the growing fireball of flames. Then it stabilized with its nose forward, which was exactly the wrong orientation, since it was that part of the capsule's skin that was the thinnest. There was only an inch of insulation in the top area, compared to the 6 inches along the bottom. And 3 inches of that was expected to burn away during a normal reentry.

Radio contact with the ground was lost. He heard and felt the explosions of the *Equipment Module*'s overheated fuel tanks, and from his seat, he watched the overhead exit hatch bulge inward under the head-on blast of air. The rubber seal on the hatch began to smoke.

As his cabin walls began searing, Volynov watched as smoke from the singed insulation filled his *Descent Module*. Since he didn't have a space suit to wear, he could feel the heat against his unprotected skin. His body strained upward against restraining straps as it tried to fall "down" onto the steaming hatch. Concluding that he only had seconds to live, he grabbed his logbook, tore out the most recent pages, and stuffed them deep inside his jacket. By some chance, he thought, they might escape full incineration.

Then there was a crash, and his module tumbled violently. Miraculously, it then settled itself into the proper orientation, its shielded bottom finally pointing in the correct direction, facing the superheated plasma shock wave. His body sagged backward into his couch. Whatever mechanical failure had prevented the separation had itself been overcome by the severe stresses of reentry. All of this was taking place during the longest minute of Volynov's life.

His next worries concerned landing and recovery. Had the flimsy parachutes in the outer containers been damaged by the flames? The capsule was spinning rapidly, as the gas for its control thrusters, which were supposed to stabilize it, was exhausted. Would this tangle the lines of the parachute? Where was he going to land, and would his recovery beacon work and guide the searchers to him?

Two thousand kilometers short of his aim point (and the waiting rescuers), Volynov rode the *Soyuz* down onto the snowy Ural Mountains near Orenburg, Russia (normal landings were in north central Kazakhstan). The parachute worked partially, although the spin of the capsule did wind up the shroud lines and partially close the canopy. Then he reached the ground.

The impact force tore him from his seat and threw him across the cabin, knocking out several of his upper front teeth. He tasted blood filling his mouth. But in the silence after landing, he could hear the hissing of his overheated capsule lying deep in the snow. It had cushioned his impact just barely enough, and he knew that he had survived, so far. It was then that he felt the bitter cold seeping into the ship—it was −38 degrees Celsius outside (about −40 degrees Fahrenheit).

Ground searchers did not know that he was alive, although radar did give some indication that the capsule was far off course. Many hours later, helicopters spotted the capsule's parachute and landed nearby. The rescuers were unsure whether they were on a rescue mission or whether they would simply recover a dead body. They found the capsule's hatch open, nobody inside, and no trace of the cosmonaut.

Volynov had quickly figured out that he would die in the midwinter cold if he stayed where he was, so he set out on foot toward a distant vertical line of smoke in the sky. It was just before noon when he landed, and the weather was clear. He found the hut of some peasants only a few kilometers away, and they took him in and kept him warm until searchers followed his footprints—and the bloody spots where he had been spitting—in the snow.

Naturally, no news of this was ever printed in the Soviet press. This kind of information was kept strictly secret. Even years later, the post-

Soviet Russian space program kept the information to itself until 1997, when an official new history book briefly mentioned the incident, clearing the way for a few newspaper accounts. But the incident had taken place long before and the problem that caused it had long since been repaired, so the Russians saw no need to worry NASA with the story.

More serious for current NASA plans was the Russians' continuing reluctance to provide information about more recent problems with the *Soyuz*, especially in the official reports concerning its role in the *International Space Station*. One such incident that occurred in 1988 was the subject of an independent study that I was asked to carry out in 1997. This was a special consulting contract for NASA's safety office, and I was being paid by a contractor, not by NASA. This time it wasn't a question of ancient space history. It was a more recent problem with the adequacy of the Russians' candor about potential problems with the *Soyuz*.

In September 1988, the *Soyuz TM-5* was headed back to Earth. It was commanded by veteran cosmonaut Vladimir Lyakhov, but the copilot was an Afghan man. He had been given six months of hurried space training and then sent up as a publicity stunt to encourage the pro-Soviet puppet regime in Kabul in its losing war against rebels. The entire space mission had been thrown together quickly, without thorough planning or training. The Russians knew that the Afghan was only along for show, and they expected him to be on his best behavior.

The spaceship's autopilot counted down to the rocket burn that would send it back to Earth. Shortly before the scheduled burn, the autopilot "armed" the rocket engine and prepared to send the "fire" command. Just seconds before ignition, however, infrared horizon sensors on the spaceship became confused by the setting Sun, and an "ignition inhibit" command was generated. The "fire" command had also just been issued, however, and the two commands canceled each other out.

The cosmonauts spent several minutes puzzling over the absence of the rocket burn. Meanwhile, the Sun had fully set and the horizon sensors were no longer in disagreement. The "inhibit" command was removed, and the "fire" command unexpectedly went into effect. The engines lit, startling the cosmonauts, who manually shut them down within a few seconds. Had the burn been completed after such a delay, the capsule would have splashed down in the northern Pacific Ocean, far from any possible rescuers.

After two additional orbits around Earth and intensive consultations with Mission Control, the crew was ready to try again. They recon-
figured their navigation system to ignore the horizon sensors entirely.

But seconds after starting the planned 230-second-long rocket
burn to head back to Earth, the spaceship's computer detected a malfunction in a navigation instrument. It shut off the engine again, halted
the automatic sequence, and sounded an alarm.

Mission Control Center director Viktor Blagov later explained that
"it is clear that a link between programs came into operation in the computer that had not been envisaged by the programmers." In English,
there was a software command error. It originated in the rushed and
careless flight plan preparation.

Lyakhov quickly determined that the spaceship's orientation was
proper, and he assumed that the shutdown was simply a leftover problem with the navigation device. He ordered the computer to resume
the program for descent. The rocket engine fired a second time. But
then, 49 seconds later, the computer again detected an error and shut
off the rocket engine for a third time.

"When the engine cut itself off for the second time, I felt very much
like starting to descend, and I restarted it," Lyakhov later told a press conference. "Naturally, I was conscious of the fact that the restart might not
be without consequence. That was indeed my mistake, and I admit it,
but there was very little time then, merely seconds, to make the decision. And I also wanted to land very much, and it was already the second
attempt, after all."

At this point, when the engine stopped for the third time, Lyakhov
did what according to Blagov "he should have done from the outset, curtail dynamic operations and await communication with the control center in order to report on the situation." Mission Control would provide
new instructions in 10 or 20 minutes, and the landing could resume on
the next pass.

But the Afghan pilot, Abdul-Ahad Mohmand, had a different idea.
He knew from his own combat flying experience that the first thing a
pilot in a contingency situation must do is to assess the status of his vehi-
cle. While Lyakhov relaxed and waited for radio contact with Moscow,
Mohmand ran his eyes over the control panel and its gauges and timers.
And then he nearly shouted with alarm.

Although the computer had shut off the rocket burn the last time the burn was aborted, it was still following Lyakhov's command to ignore its earlier automatic shutoff. Thus it was continuing with the normal descent sequence of the other commands. It had already concluded (falsely, of course) that the ship was descending toward Earth after a completely successful full-duration burn. This meant that after it had received the expected cue of "rocket shutdown," it knew that the next step was to fire the explosive bolts to separate the *Soyuz*'s *Descent Module* from its *Equipment Module*.

But the ship was still in orbit, and the rocket engine in the *Equipment Module* was the only way to get home. Mohmand saw that the clock showed less than a minute to the firing of the explosive separation bolts!

Shocked into action, Lyakhov was able to shut down the sequence with another command. If he had waited for advice from Moscow, and if the Afghan had obeyed his instructions to sit quietly and trust in the Russian commander, the *Soyuz* modules would have separated while the spaceship was still in orbit. The two men would have had battery power and air for only a few hours, and then both would have died.

"By his timely actions he prevented very serious trouble that might have happened in orbit," a veteran cosmonaut explained later, referring to Lyakhov's button pushing. "[He] shut the program down and averted—well, let us not speak of it at this stage—but possibly very serious trouble in which the crew might have found itself in orbit." The euphemism "serious trouble" refers to their suffocating to death.

Yet a decade later, the official Russian *Soyuz* flight safety report to NASA mentioned only the *Soyuz TM-5*'s delay in landing (the crew was able to complete the descent the following day). No mention was made of the confused computer that had brought the men to within seconds of death. The story was only vaguely hinted at in public, but it was common knowledge among the cosmonauts, who told their European colleagues the whole story. Mohmand, too, described the incident in detail to an interviewer in Germany, where he was living in exile following the collapse of the Soviet-backed Afghan regime that he had served. He now works as a laborer in a print shop, where his coworkers humor his wild tales of space flight.

"But I had performed well, had made a good flight and had overcome difficulties," he had said. "I had assisted my commander during the difficulties. After the engine had stopped, I had said that we should first try to find out what was going on, because at the time, we didn't have contact with [Moscow]. I told the commander once, twice, three times. Later, on Earth, I heard that we had had only twenty seconds before the vehicle would have separated. The second time, we even had less—only two seconds. I did all that, so after the landing, I was happy."

I arrived at a preliminary reconstruction of this near-disaster based on interviews that were vague when taken one by one but that when taken together painted an alarming picture.

Independent space journalists in Moscow were also working on this story. On the tenth anniversary of the incident, in mid-1998, the biweekly space magazine *Novosti Kosmonavtiki* *(News of Cosmonautics)* printed an article on the incident. A space engineer named Aleksandr Fyodorov provided his own version of the events. He reported that when the automated jettison sequence advanced to within two minutes of execution, a warning tone sounded and an indicator light came on. Fyodorov states that it was these signs, and not Mohmand's cries, that warned Lyakhov, the Russian pilot. At least, that was Lyakhov's story, and he was sticking to it. Other *Soyuz* cosmonauts I've talked to, however, are unaware of any such alarms.

At this point, Fyodorov continued, Lyakhov called the control center by radio and requested permission to manually issue a "stop" command and switch some sensors off. He waited a full minute or more but received no answer from Moscow. The article makes no mention of any of Mohmand's warnings.

Finally, Lyakhov stopped waiting and entered the commands on his control panel, stopping the sequence. In this account, the clock had reached 79 seconds before jettison, which would have spelled certain doom for both men. Fyodorov quoted Lyakhov as saying that only after the sequence had been aborted did the crew realize the depth of their peril in those critical seconds.

They had to spend an extra day in space, wearing protective space suits designed for only a few hours. They had emergency rations, but they had nothing to do except sit in their seats and think about Earth. Finally, their orbit brought them back across Russia.

On Moscow TV, cosmonaut Vladimir Dzhanibekov was asked how the foul-up could have happened. "Well, evidently, a kind of blunting of vigilance, frequent in success, plays its part here," he explained. "Some caution and vigilance was lost. Perhaps there was a little less attention, a little less concentration." Vladimir Semyachkin, a senior operations specialist at Mission Control, was blunter: "The main cause of the crew's daylong torments was acknowledged to be a combination of incorrect actions of the crew commander and mission control personnel." I've worked with Semyachkin and trust his judgment, since he struck me as a ferociously candid "take-no-prisoners" kind of operator.

Whichever version of the crisis is closer to the truth, one alarming aspect stands out: The computer foul-ups nearly killed two men, and they would have worried any potential commercial clients interested in using the *Soyuz* capsules. So the Russians quietly fixed the computer problem and never bothered to tell NASA about the incident.

My private "Soyuz Landing Historical Reliability Study" was delivered to NASA on March 19, 1997. My overall task had been to assess the flight history of *Soyuz*-type space vehicles, as far as it was known from public news releases and private sources, in order "to ascertain the demonstrated reliability of key functions associated with return to Earth from a space station." In particular, I was supposed to watch for flight events that could be relevant to the success of a medical evacuation mission, which was the most likely scenario for the use of a space station "crew rescue vehicle."

I put together a data base that recorded 222 flight sequences from 186 separate *Soyuz*-type spacecraft, some with cosmonauts and some on autopilot, and all of the known failures during these sequences. As often happens in a highly challenging space engineering program, I found that the failures were heavily front-end loaded in terms of chronology (they mostly happened early in the program). In the 173 undockings, 2 failures occurred, the last in 1976. In the 97 module separations (separation of the *Descent Module* from other modules of the same spacecraft), 1 failure occurred, in 1969. In the 110 spacecraft landings (including reentry and parachute deployment), 8 failures occurred, all but one of them in the first five years of flight operations (the eighth and last occurred in 1980). In contrast, the rate of deorbit burn failures (8 out of 182 attempts, or 4 percent) was fairly constant across the three decades. This was probably due in large part to operator error as well as the rate of hardware breakdown.

It was a challenge to determine the completeness of the data base, because so many of the incidents that did get reported did so only by accident. For example, *Soyuz* spacecraft often wind up on their sides after landing, which is a serious concern if it is necessary to evacuate an injured or unconscious crew member. Documentation of how often this happens is inadequate. Accidents are rarely reported in the Russian media, and most information about such events comes either from inspection of video and photo records made at the landing site or from the personal testimony of crew members. I warned NASA that it should presume that a large number of similar incidents that were not photographed and not described orally have also occurred.

I found that the official Russian reports on *Soyuz* flight events were inadequate as a means of assessing the effectiveness of medevac (the medical evacuation of sick or injured crew members) missions. As an example, I compared the RSC-Energia report NASW-4727 to NASA on the anomalies in the *Soyuz TM-2* through *Soyuz TM-15* missions with the anomalies registered in my study. The Russians reported only trivial procedural deviations, while the flight crews involved in the missions confronted serious operational difficulties, sometimes life-threatening, which would have called the success of a medevac mission into doubt.

The incidence of "hard landings" appeared to actually increase with time, over the history of the program. However, I suggested that this was probably an artifact of the data collection process. Although official accounts continued to omit any such information, nongovernment news media representatives have covered the more recent landings more thoroughly.

As for the hard landing in 1980 (caused by the failure of the soft landing engines), as described in Chapter 8. Russian files released in the mid-1990s show that they considered it to be one of the four worst problems they had ever had with that generation of *Soyuz* vehicle (used between 1972 and 1980). Then, after my study was completed, another *Soyuz* made a similar "crash landing." NASA, as far as I could tell, wasn't alarmed.

My NASA contract monitor presented my results to the *ISS* Independent Assessment group in April 1997. He told me that the reaction had been very positive, that the report should be presented to *ISS* program management and probably also to headquarters.

But after several weeks of waiting without any response, he called his contact on the group, Hugh Baker. Baker told him that "it was not necessary to present the results to anyone else." The reasons were that the *Soyuz* would someday be replaced by the *X-38*, that other people were studying the problem, and that "our analysis confirmed what [his office] knew about the *Soyuz*, anyway." Six months later, after I had mentioned the study in my congressional testimony, my monitor received an urgent call from the Phase 1 office, asking for a copy. But neither they nor anyone else ever called back to comment on it.

There's one last "explosive secret" about the *Soyuz*, and it deals with a particular item carried inside it. When the Russians were invited to take part in the *International Space Station*, the invitation was given with open arms. Nobody expected them to come armed, but they did. Russian participation means that there are guns on board the *ISS*, and the guns belong to the Russians.

This is not quite as alarming as it sounds, and officially it's no secret. However, I could never find any mention of this design feature on NASA Web sites or mission press kits. Actually, it's a safety feature, and not an unreasonable one.

American astronauts who trained for the 1995–1997 *Mir* visits, and later as part of the *Soyuz* spacecraft crews for the *International Space Station*, encountered a unique feature that cosmonauts need to master: target practice. They have to know how to load, aim, and fire the special survival gun that has been on board all *Soyuz* spacecraft throughout their 30-year history.

The triple-barreled gun can fire flares, shotgun shells, or rifle bullets, depending on how it's loaded. The gun and about 10 rounds for each barrel are carried in a triangle-shaped survival canister stowed next to the commander's couch. The gun's shoulder stock opens up into a machete for chopping firewood.

Familiarization with the gun usually takes place during survival training at the Black Sea, when the crews train to safely exit a spacecraft floating on the water (although a firing range at the cosmonaut center at Star City near Moscow is sometimes used for training). After floating around in the water for a day or two, the astronauts and cosmonauts take a few hours to fire several rounds from each chamber off the deck of the training ship.

"It was amazing how many wine, beer, and vodka bottles the crew of the ship could come up with for us to shoot at," astronaut Jim Voss told me. "It was very accurate," he continued. "We threw the bottles as far as possible, probably 20 or 30 meters, then shot them. It was trivial to hit the bottles with the shotgun shells, and relatively easy to hit them with the rifle bullets on the first shot."

"It is a wonderful gun," agreed *Mir* veteran Dave Wolf. "I found it to be well balanced, highly accurate, and convenient to use."

Mike Foale trained with the gun and found it to be pretty standard. "Other than firing flares, bird shot, and a hard slug from its three barrels, during sea and winter survival training, I can't say it is very unique," he told me. He added, as if in reassurance, "The *Soyuz* commander controls its use."

Every *Soyuz* spacecraft carries such a gun, although none of these guns have ever been unpacked in flight. And they have never been needed, with the exception of an incident in 1965, when bears (or wolves—the story varies) chased two far-off-course cosmonauts. The guns are often presented to crew members as postflight souvenirs. Although several survival kit bags have shown up at space auctions, I've never seen any of the guns for sale.

The *Soyuz* spacecraft will continue to be crucial to *ISS* for years to come, until NASA or another international partner develops an alternative means for the station crew to return to Earth when space shuttles aren't present. *Soyuz* has overcome a variety of problems, and it continues to evolve, with a new generation of vehicles showing up every decade or so. The model used on *Mir* and the early *ISS* missions, *Soyuz-TM*, is to be replaced in 2002 by the *Soyuz-TMA* design, which has larger seats (for taller passengers) and simpler controls.

With American lives at stake, NASA has every reason to want to know everything about the spacecraft's past, present, and future. Recent experience has shown that they won't succeed without vigorous, even confrontational, interfacing with Russian space officials, who so far have held most of the cards—as well as the only gun.

12

ISS in the Wilderness

"We need to assume that "the Big We"—the President, the Congress, the NASA, the aerospace industry, the public—right now cannot manage any space program. Think about it: We have spent $20 billion over 14 years for a space station and we do not yet have a screw in orbit!"

Astronaut Story Musgrave, 10/20/98

"True believers do not automatically abandon their cause when reality intrudes in discomforting ways. They rarely admit they were wrong or change their behavior, especially when they keep meeting other people who share their fantastic beliefs. Believers so sustained often seek new beliefs to validate old behavior and new interpretations to explain why prophecies fail to be fulfilled."

Howard McCurdy, *Space and the American Imagination*, 1997

In 1997, at a private team briefing on the latest inadequacies in Russian space performance, *ISS* program manager Randy Brinkley sounded almost plaintive. "The Russians have just *got* to stop surprising us," he lamented. Even in public, Brinkley grudgingly admitted to a reporter, "I guess maybe we could criticize ourselves for being overly optimistic five years ago in terms of predicting some of the difficulties that we would experience."

That same year, Lynn Cline, NASA's associate deputy administrator for external relations, told *ABC News*, "We never could have anticipated that things would be as bad in Russia as they have been," adding that "the relationship has been very difficult." It had been Cline's job to anticipate exactly such difficulties, but like Brinkley, he blamed the Russians for surprising NASA.

By the time the shuttle-*Mir* program was over the following year, NASA leadership was having a harder and harder time rationalizing the apparent collapse of many of Russia's promises. But for the first five years, their faith had been strong.

"I am very encouraged by what we saw in Russia and what we heard at the [Critical Design Review]," Brinkley had told space workers in 1996. He dismissed any skepticism: "I've heard continued predictions of doom and gloom, but we heard that four years ago." He granted that 1997 and 1998 would be "very critical years," but he promised that things would get better because "we have lots of reserves." That is, NASA had set aside spare money and spare time to accommodate any unexpected problems that might arise.

Brinkley went on to describe the latest sincere promises for funding that he had been given in Moscow. As a result, he explained, "we feel a lot more confident in their ability to meet a December 1998 launch [of the *Service Module*]. We see clearly a different attitude from the Russians."

Wilbur Trafton, a former U.S. Navy aircraft carrier officer who worked directly for Dan Goldin at NASA headquarters, repeated the official optimism. "At this time we expect to stay within our [budget] boundaries and launch on time," he had proclaimed in July 1996. "We have an executable schedule with costs that maintain acceptable reserves within our budget cap," he added. As to widespread worries about whether the Russians would carry out their end of the bargain, Trafton remained confident: "I would point out to you that Russia, even in the darkest days of the Cold War, always honored their international agreements."

White House officials echoed these sentiments. In a grand example of governmental doublespeak, Jacob Lew, director of the Office of

Management and Budget, told Congress on August 26, 1998, "The best information we have at the moment does not lead us to make the judgment that we need to jump to the conclusion that the Russians will not be able to meet any of their commitments." The English of this is that NASA and the White House saw what they wanted to see.

But by the beginning of 1998, reality had intruded on even the most die-hard optimists at NASA. Although the U.S.-funded *FGB* module was still slated for flight at the end of 1998, the next step was the Russian-financed *Service Module*, and its launch date was rapidly "moving to the right." Another spacecraft, the heavy robot space freighter that would replace the *Progress* robot supply ships, had been promised as part of the deal that authorized the last two shuttle-*Mir* dockings. But it had totally disappeared, no more substantial than the paper envelopes that its design had been drawn on. Later Russian modules, for science, stabilization, and power, were also quickly fading into the distant future.

My friends within the *ISS* project passed on accounts of growing desperation and despondency. There was also a grim determination to keep the bad news quiet as long as possible. Meetings were called without written announcements or published minutes. Topics were verbally discussed, with the admonition "no e-mail mention." Decisions were made "off the record," deliberately evading the "paper trail" of rationale and responsibility. The project looked to be lost in the wilderness.

As budgets dwindled, Russian space officials also began to cut back on the traditional safety measures for *Mir* operations. Some steps were simple, such as sending the primary and backup crews from Moscow to the Baykonur launch site on the same airplane (as a precaution, they had previously been sent on separate planes). Others were more significant, such as ending the practice of delivering each *Soyuz* capsule to Baykonur one cycle early, so that it would be immediately available in the event of the need for a rescue mission. For years, the Russians had trained veteran pilots to be "rescue commanders." These men would fly solo to pick up the two cosmonauts who were in danger on the space station. The Soviets had viewed this as a prudent backup plan, but by the mid-1990s, the Russians discovered that they couldn't afford it.

One already existing Russian space system was quickly collapsing, and there was no replacement in view. This was a communications radio relay system, and it was vital to early *ISS* operations. It included two satellites in 24-hour orbits over the Atlantic and Indian Oceans. They would provide critical voice and data contact with the space station crews.

The satellites were called *Altair-1* and *Altair-2*, and the whole system was called *Luch*, or "Ray." It was still in use during the shuttle-*Mir* program, but since it was owned by the Russian Defense Ministry, the civilian space program had to pay cash, by the minute, for its services. Nobody ever seems to have determined how much that cost, but private NASA estimates put it at thousands, if not tens of thousands, of dollars per minute.

Using these satellites, cosmonauts on *Mir* could communicate with Earth every time they swung around to its eastern hemisphere, even when the orbits in which the station would spend half of each day were too far to the south to be within range of the in-country ground sites. And there was one additional complication: Often *Mir*'s antenna wouldn't work because its control electronics would quickly overheat when the station's internal cooling system broke down, which it did periodically. But the capability was available, at least for short periods, when it was urgently needed.

Without this relay, Russian satellites could contact Mission Control only during the times when they were passing over Russian territory. There was a string of tracking sites running from St. Petersburg to Moscow to Baykonur, then across Siberia to Vladivostok and finally to the Kamchatka Peninsula (although that station was often broken). In better days, the Soviets had sent tracking ships out into the Atlantic and the Pacific, but under the deteriorating economic conditions, the majority of the ships had been sold as scrap.

Then the last *Altair* broke down. The Russians promised to replace it, but they didn't. A communications satellite institute in Siberia built a replacement, but wouldn't turn it over until it was paid for in full, in cash. Nor was there any money for a *Proton* launch vehicle.

The Russian Space Agency had promised to have an *Altair* satellite operational for early operations with the *Service Module*. It would be needed until the American/Russian crew became able to communicate with the fleet of NASA relay satellites using the additional equipment that was due to arrive a year after the *Service Module*. When it became clear that the promise was not going to be kept, NASA officials began a crash project to create an "Early Comm" radio link. This system, which was quickly developed using some classic NASA ingenuity, gave the station a low-quality but round-the-clock voice link through the NASA satellite network.

However, it wasn't free. And as with so many other replacements for failed Russian promises, the mounting costs finally got management's attention.

"Time and again, we were told that the problem would be resolved," Goldin finally admitted to Congress in 1998, not having to add that he had believed the Russian promises every time. "Last summer, Russian government officials said the lack of funding would be resolved—it didn't happen." In a September 21, 1998, *Washington Post* story titled "NASA Seeks to Bail Out Russian Space Agency," headquarters official Joe Rothenberg stated, "A year ago, we wouldn't have predicted things would be this bad." Admitting that he was "frustrated and angry" over the Russian failure to fund the *Service Module*, Goldin told a congressional committee, "We were too naïve in expecting them to act like we act."

NASA demonstrated desperate spin control on the bad news: "I think the cooperation has gone very well within the sphere of the possible," NASA's chief liaison officer in Russia, Doug Englund, told a Reuters reporter. "There were a lot of decisions taken probably by both sides in anticipation of economic developments that didn't eventuate. Now we are all having to deal with the realities of the Russians' economy that didn't move the way we all hoped and predicted and projected that it would" (Reuters, November 16, 1998). This gobbledygook (I don't recall ever seeing the word *eventuate* before in my life) was only a smokescreen to deflect the blame for getting blindsided by what outside observers had always warned was inevitable.

Although White House officials also, in turn, admitted error, they stuck to the line that the problems were unavoidable. Congressman James Sensenbrenner sent Deputy Secretary of State Strobe Talbott a pointed question: "This proposal made specific claims to cost savings. Were these expectations too high based on Russia's economic condition?" Talbott responded by making excuses: "There was no way in 1993 that NASA or the Administration could have anticipated the current economic crisis in Russia."

By 1998, administration officials stopped claiming that the partnership would save money and time. Late in 1998, President Clinton told reporters: "If we were required now to help the Russians during this difficult period—which will not last forever—so that they could continue to participate, I would be in favor of that. I think that it's very important that we have the Europeans, the Japanese, the Canadians, and the Russians in the space station venture."

Space officials echoed that theme. "It is a political program as well as a science program," said Douglas Stone, program manager at Boeing. "Just getting these countries to work together peacefully is worth more than the price of the station."

Partnership proponents also advised against second-guessing the decisions that were turning out to have been ill advised. "Deploring in 1998 the consequences of assigning Russia a central role in the station program, as occurred in 1993, really is not very productive," scolded John Logsdon, who specialized in international cooperation studies at George Washington University. "Russia might not have accepted any lesser role, given its space accomplishments. If wanting Russia in the station program was justified by political and security imperatives, giving it a critical role might have been a necessary price."

In November 1997, Logsdon also suggested to *Space News* that since all of the obvious reasons for the partnership had failed, perhaps there were some *secret* aspects that the public didn't know about that *were* working. That would explain why the White House remained enthusiastic about the partnership. "None of us on the outside knows the full range of the US-Russian bargain," he speculated, "in terms of what the quid pro quos were when the decision was made to bring the Russians into the station. When you see how strongly NASA and the White House push to keep the partnership going, it suggests that in their judgment this is a foreign policy and security policy success."

John Pike, the space expert for the Washington-based left-wing lobbying group called the Federation of American Scientists, was unperturbed about the delays and cost overruns. "It's basically like getting some repair work done on your house," he explained. "Once the project is started, there's no way you can stop it until the house is finally completed. And I think we're in the same situation with Russia."

He, too, had settled on a new rationale for the project once the original justifications had collapsed. "It's a lot cheaper for the United States to pay to keep those rocket scientists in Russia," he argued in the classic false dilemma, "than for us to have to pick up the pieces if they all move to North Korea or Iran."

In a PBS radio discussion with me a few years later, Pike elaborated on why politics was the *only* way in which the station could ever have been funded. "I would certainly hope that it is a political project," he said provocatively, "because it is the politicians and the voters and the taxpayers that are paying for it. It has to make sense politically, and human space flight has always been about politics. Yuri Gagarin, the first Soviet in space, was about politics. John Glenn, American hero, was about politics. Neil and Buzz on the moon was about politics. During the cold war it was about racing the Soviets and now it's about cooperating with the Russians. If the human space flight program does not make sense politically, it's not going to get funded and it's not going to happen."

Although NASA and space enthusiasts in general have always been devoted to space flight, Pike stressed that this wasn't enough. "Historically, since the beginning of the space age, piloted space flight has been about the Russians," he explained. "The space station got into trouble under the [first] Bush administration because they couldn't explain how the project related to the Russians, Reagan knew that he needed the station to beat the Soviets, Bill Clinton knows that he needs it to cooperate with the Russians, and as long as the political equation works, I think the station is going to be safe. If the politics comes unraveled, though, the space station, and I think our entire space program, is going to be in real danger."

A shuttle astronaut who had been assigned to NASA headquarters in 1997–1998 spoke with me at length in April 1998, not for attribution. There were many conflicting views within NASA headquarters, he revealed. Some officials had always been skeptical about whether the Russians would pull it off. Others suspected that the only reason the Russians had joined was to damage or destroy the American program. And there were some who sincerely believed that the Russians would be there when they said they would. "When you immerse an operational program in politics," he observed, "it is not without its challenges."

He put the project into its proper political perspective. "There are priorities that the White House has in dealing with the Russians," he explained, "and the space station is one of those pawns, one of those tools." The partnership was never intended to be good for the space program, but it was supposed to be good for U.S.-Russian relations. As for giving the Russians a critical part to play, he agreed with Logsdon's psychological assessment: "One may say this was very foolish, or it might have been necessary for them to keep their esteem."

"People who make the decisions, at the highest levels of the U.S. government, believe that without the Russians, we won't have any more big space projects," my source explained. "They believe that Congress would never support a human piloted Mars mission without the Russians, and that the whole world has come to the belief that the only way to fly to Mars is as an international project with the Russians playing a major role." This echoed a comment that Brinkley had made at an in-house "all-hands" meeting in Houston in May 1998: "*ISS* must be successful before we can conduct deep space exploration."

The delays created anguish and frustration at all levels of the space station team, but it turned out that these delays were not without benefit. They allowed space engineers in both countries to concentrate on several significant problems that would not have been fixed if the station had been launched on time. In its original schedules, NASA had failed to allocate adequate time for testing and adjustment of the hardware. Now it had the time.

The extra time waiting for launch was put to good use, and two years later, after the initial modules had been hooked up with astonishing ease, senior program managers told me that they believed that the lack of trouble in 2000–2001 was due to the extra testing conducted in 1998–1999. NASA's original delusion had been that the agency could build it right the first time, then shake it down in orbit when nobody was watching. Casually dubbed "ship and shoot," the strategy was a money-saving gimmick designed to appeal to Goldin's passion for "faster, better, cheaper."

It would have been a disaster. "The original idea was totally bogus," a senior NASA station manager and former astronaut confided to me in April 2001. "We'd have had all the software problems in orbit instead of during integrated testing at the Cape," another senior engineering manager—an old friend of mine from the early space shuttle days—told me in the parking lot at the Johnson Space Center. Center director George Abbey was already gone, and Goldin's days were numbered, but the instinct to self-preservation was still strong: Both of my sources explicitly asked that their names not be used.

Other watchdog groups had used the delay time to identify safety concerns. While waiting for assembly to begin, specialists for the Government Accounting Office studied NASA safety documents concerning the *Service Module*, which was being built in Russia. In congressional testimony, they described in detail four "significant areas of noncompliance" with NASA safety requirements.

The most serious violation was the high level of noise inside the Russian modules. Fans and motors would be running constantly, and the noise this would create would be far in excess of the allowable levels set by the U.S. Occupational Safety and Health Administration. The noise would be enough to make normal crew conversations impossible. It might also mask the warning tones that would sound during emergencies. Medical experts worried that the noise could disturb the crew's sleep and lead to psychological problems. In some cases, such noise levels could result in permanent hearing loss.

Other safety issues involved protecting the modules from penetration by space debris, verifying that the windows were strong enough to withstand years of space exposure, and designing the equipment so that it could function even when air was leaking out of the station.

Allen Li, the associate director for the GAO's National Security and International Affairs Division, told the subcommittee that "the lack of approval [of Russia's request for waivers of these requirements] currently stands in the way of the *Service Module* launch." Despite Russia's noncompliance, there was never any possibility that NASA would use these issues to force a delay in the launch. Instead, everyone expected NASA to issue waivers for the violations, based on Russian promises to fix the problems at some later date. This was in spite of a poor record of Russian follow-through on previous safety waivers from NASA.

"When you're up against the stops, you have no other options," former astronaut Blaine Hammond told me. Hammond, who had been the safety representative from the Astronaut Office, left NASA in 1997 over disputes about the safety of U.S. visits to the *Mir* space station.

Because of the persisting noise problem, space station crew members were expected to wear protective headsets to deaden the harmful noise levels that they would be exposed to. Hammond called that solution "just ridiculous." He described headsets as "inconvenient and really uncomfortable," and he predicted that most of the crew members would simply not use them. "You can mandate it," he told me, "but then they ignore it."

Li had cited a study of 50 cosmonauts who had served aboard *Mir*, which was just as noisy as the new space station modules. "Virtually all suffered temporary hearing damage," he said, "and some had permanent damage that disqualified them from future space flights.

"At least one NASA astronaut who stayed aboard *Mir* for an extended time suffered significant temporary hearing loss," he continued. In response to a flippant comment from a cosmonaut that "you can get used to it," a veteran NASA doctor-astronaut told me, "It's not a matter of getting accustomed to the noise, it's a matter of physical ear damage."

"Don't blame this one entirely on the Russians," another astronaut e-mailed me confidentially in February 2000. "Our modules and those contributed by the international partners will be equally, if not more, noisy," he warned. "Most racks have a fan that is about as noisy as it can get, and even one running in a medium sized room is deafening," he continued. "Having dozens of things spinning at the same time along with all the other pumps and air handlers will be excruciatingly painful. Permanent hearing loss is inevitable and I think [NASA] is unwilling to do anything about it.

"The requirements were there for maximum noise levels," he explained, "but no one took them seriously. Even early on, folks knew the hardware was noisy but everyone just seemed to hope that the problem would go away while they were busy working other 'important issues.' Of course it's too late to change anything now." Just as he predicted, when an air pump in the newly installed *Airlock Module* was first activated in July 2001, the shriek was so excruciating that astronauts had to flee from the module and the one next to it.

As part of Li's presentation to Congress prior to the launch of the *Service Module*, he delivered an 11-page written statement entitled "Space Station: Russian Compliance with Safety Requirements." It dealt with specific features of Russia's *Service Module* and of the *FGB* module, built by Russia and launched in November 1998 as the cornerstone of the *ISS*.

"*[FGB]* and the *Service Module* still do not meet some important requirements," Li reported. "These shortcomings increase the risk of health hazards and of what NASA terms 'catastrophic' failure of the modules."

Li described how NASA explained these shortcomings to GAO investigators: "NASA officials said that shortfalls in Russian funding, designs based on existing Russian hardware, and technical disagreements with Russian engineers are the main reasons these modules do not comply with safety requirements," he said.

As an example of how NASA had granted waivers on the basis of promised future repairs, Li discussed the issue of shielding the station from collisions with space junk. The *Service Module* would be equipped with additional shields over a four-year period after its launch. While Li admitted that "this still does not meet the original requirement," the Russians had prevailed when they pointed out that adding the required shielding at launch would make their module too heavy to fly.

NASA also approved a waiver for the *FGB* module on the issue of its ability to operate in the event of an air leak. A leak could be caused by a collision with space junk or by the failure of a hatch or an airtight seal. Since the module's computers are air-cooled, a loss of air would result in rapid overheating and breakdown of their circuits. This would result in a loss of control of the station.

NASA accepted this risk in August 1998, on the expectation that the period of exposure to the risk would be short. At that time, the *Service Module* was planned to link with the *FGB* five months after its launch and take over all of the control functions. Subsequent delays for the *Service Module* meant that NASA's waiver was granted on an incorrect estimate of the duration of the risky situation. The *FGB* flew without the *Service Module* for almost two years.

Li provided another example of a situation in which NASA granted a safety waiver based on assumptions that turned out to be faulty. "NASA approves noncompliance with safety requirements when it determines the risks are acceptable because plans are in place to mitigate risk," he explained, "or the deficiencies will last only a limited time."

NASA's specification for space station noise stated that it could not exceed an average of 55 decibels (dB) over a 24-hour period. For the Russians, this level was quickly bumped up to 60 dB. Li explained that this was "because Russian space officials would not agree to meet the general specification. This new value is usually compared to a typical 'noisy office.'"

In December 1998, a visiting space shuttle crew measured noise levels of between 65 and 74 dB inside the *FGB* module. The working comparisons for these levels are that 70 dB is equivalent to the sound level of normal road traffic, and 80 dB is equivalent to the level of a subway or of rock music. On a subsequent shuttle visit to the space station in May 1999, astronauts installed mufflers and other noise reduction devices. They then measured average noise levels in the 62- to 64-dB range.

Li described in detail what GAO investigators had identified as "potential problems with deferring corrections until after launch." In addition to unexpected delays in beginning the repairs, Li gave examples of how fixing things in orbit can turn out to be much more difficult than expected. The visiting shuttle crew in May 1999 encountered significant difficulties when they attempted to install equipment to reduce noise levels in the *FGB* module. Li quoted NASA officials as saying that "the crew did not receive adequate training and instructions on how to install the mufflers, and could not get them to fit properly."

"In attempting to force the mufflers around the ducts," Li continued, "the crew crimped the ducts." More repairs were then needed for the equipment that was damaged during the initial attempts at repair.

Regarding the expected excessive noise levels on the then-still-unlaunched *Service Module*, Li told Congress that "NASA officials believe that installing some noise reduction devices will be difficult in orbit, and have suggested to their Russian counterparts that installation be done on the ground." But the Russians ignored these suggestions, and, as they expected, NASA did nothing about it.

The *ISS* partnership also suffered from another lamentable legacy of the shuttle-*Mir* program—"mission creep." This is an unexpected expansion of one side's promises and duties with respect to the other side, and what this process did to expected project costs. The United States started out as a paying customer on *Mir*, just like all the previous non-Russian users of the space station. The United States agreed to provide specified monetary payments and services, and in return, the Russians agreed to provide specified facilities and support. The shuttle-*Mir* program was to be the practice round for the next step, the genuine partnership (as spelled out explicitly in intergovernment agreements): the *International Space Station*.

But what actually happened was not predicted. Caught up in enthusiasm, U.S. officials began to treat the *Mir* arrangement as already a full partnership with the Russians, with many formerly unspecified obligations suddenly laid upon the U.S. side and many formal requirements removed from the Russian side. At U.S. expense, radio relay equipment was set up and operated at NASA centers to replace the collapsed Russian space communications capabilities. At U.S. expense, Russian crew members began flying on shuttle visits to *Mir*, with no quid pro quo from the Russian side. At the expense of planned American scientific research, equipment was left off one shuttle flight after another in order to make room for last-minute Russian repair tools and emergency supplies. Finally, the United States even dumped one of its fully trained crew members because—suddenly—it was no longer using *Mir* for what it was paying for. The United States had somehow become fully committed to *Mir*'s upkeep. Thus the relationship turned out to be more costly and less valuable than had been originally planned. And even though there were some lessons that were sinking in, as the Congressional Research Service's space expert Marcia Smith pointed out to AP [on February 20, 1998], "there's still a question as to whether you needed seven people to go up and learn this."

"It's very true the Russians are going through a difficult time right now," shuttle-*Mir* manager Frank Culbertson told *Space News* in 1997. "But just as we expect them to stick with us on the international space station, to make it work together, I think it's important that we as partners stick with them during difficult times with *Mir*."

As a good partner, NASA stepped in to help with Russia's financial woes. Throughout 1998, there was pressure on space station workers to think up various projects that Russia could be "hired" to do. NASA needed an excuse for sending Russia more money. For $60 million, NASA "leased" six cubic meters of storage space aboard future Russian modules and "hired" 4000 hours of future Russian cosmonaut time in orbit. The money was supposed to be spent on making space hardware so that someday the Russians might be able to make money the traditional way.

NASA operations official James van Laak told Congress that "the Russians have no personal computers, their data is all handwritten in little green books in the pockets of operators in Moscow." At NASA's expense, the Mission Control Center in Moscow was outfitted with several dozen top-of-the-line PCs.

As time passed, the Russian government somehow continued to be unable to find the money to fund its space activities. NASA continued to buy more and more items, and circulated memos to workers asking for suggestions for anything else they wanted to buy from the Russians—hardware, training, documentation, anything they could think of—that would provide excuses for more funding. To relieve another cash crunch shortly before the first permanent crew was sent to *ISS*, NASA transferred another $35 million for more docking mechanisms, for safety equipment to be installed on Russian space suits, and for various activities such as "integration support" in Moscow. During this phase, NASA had estimated that the Russian delays were costing $3 to $5 million per *day*, so sending money to Moscow to get the Russians to speed up their efforts was a bargain.

Russia also had independent ways of making money from space. Some involved renting seats on space missions to other countries. Some involved commercial sales of rockets and other hardware. One in particular was the "icing on the cake" of Russia's financial dealings with NASA. It involved a theme from the very beginning of the formal partnership in 1993 that suddenly reappeared, once again catching NASA totally by surprise.

In 1993, the American public was told that a Russian deal to sell cryogenic (operating on super-low-temperature fuel) rocket engines to India had to be thwarted to prevent military-related missile technology from proliferating. Although many Americans were given the impression that Russia had killed the deal in return for an equivalent contract to build the *FGB*, it didn't turn out that way at all. Russia kept both deals: the original one with India (slightly modified) and the new so-called replacement deal with NASA.

The Russian rocket engine vendor, Khrunichev, had agreed not to send engine manufacturing technology to India. This was because U.S.

State Department experts were convinced that the engines could be used for military missiles, despite the fact that nobody in the world had ever used the super-cold liquid hydrogen fuel for military purposes. The engines had been built for the Soviet man-to-the-Moon program 25 years earlier, then mothballed. Instead of sending the engine's blueprints to India, Khrunichev added three more completed engines to the four already in the contract. But the value of the contract remained the same, about $200 million.

In September 1998, as NASA was struggling to invent ways to slip more money into the Russian space industry, the first of the 12KRB engines was loaded into an airplane and flown to India (the first of those engines, built into a new Indian space booster to launch communications satellites, was successfully flown into space 2$\frac{1}{2}$ years later). By the end of 1998, the Russians had money flowing in from both east and west. But just where all that money was actually going, and what good it was doing the *ISS* program, remained controversial.

13

Culture Gap

*"Before you run in double harness, look well to the
other horse."*

Ovid

*"We are inclined to believe those whom we do not
know because they have never deceived us."*
Ben Jonson, *The Idler*, 1758

As space workers at NASA struggled to understand their Russian
colleagues well enough to build a space station with them, they
often exchanged information among themselves about their experiences with Russia and the Russians. The first thing they came to realize
was that they hadn't known enough about the nature of the culture
gap between the two nations when they started work on the project.
An insightful e-mail that illustrates this point was widely circulated in
early 1997.

"Now that I've been in Russia for some time and am qualified as an
'expert,' I think it might be of some interest to you to know my general
reactions," the message began.

"Everyone will agree on the importance of collaboration with
Russia, now and in the future," the note continued. "It won't be worth a
hoot, however, unless it is based on mutual respect and made to work

both ways." The author described various ways in which the countries were interacting: "We send the Russians another millions of dollars, and they approve a visa that has been hanging fire for months. We then scratch our heads to see what other gifts we can send, and they scratch their heads to see what else they can ask for. We still meet their requests to the limit of our ability, and they limit ours to the minimum that will keep us sweet."

The message continued: "We never make a request or proposal to them that is not viewed with suspicion. They simply cannot understand giving without taking, and as a result even our giving is viewed with suspicion. Gratitude cannot be banked in Russia. Each transaction is complete in itself without regard to past favors. The party of the second part is either a shrewd trader to be admired or a sucker to be despised."

The writer seemed to like individual Russians, and he gave them credit for their successes. "The Russians have done an amazing job. One cannot help but admire their effort and the spirit with which it has been accomplished."

He then suggested various ways in which the Americans might improve their position with respect to the Russians. His tactics involved continued assistance, but only in those cases in which the Russians could prove their need for the aid: "We should insist on a quid pro quo." Further, he advised, "We should present proposals for collaboration that would be mutually beneficial, and then leave the next move to them"—that is, stop acting like a supplicant that was desperate to do favors for them. Finally, he advised, "we should stop pushing ourselves on them and make the Russian authorities come to us. We should be friendly and cooperative when they do so.

"I think there is something here worth fighting for," he concluded, "and it is simply a question of the tactics to be employed." Changing these tactics, he warned, might cause a temporary cooling of relations. "However, I feel certain that we must be tougher if we are to gain their respect and be able to work with them in the future."

NASA workers who received this e-mail were enthusiastic about its on-target assessments. Several of them e-mailed me that somebody at NASA was finally seeing the light, and that with these lessons, the rest of the team might finally be able to learn how to interact productively with the Russians across the "culture gap."

Only at the very end of the note (which I'd left off the first time I sent it out) did the irony become apparent. The letter from Moscow wasn't from some battle-scarred but wiser NASA official in the 1990s. It was dated December 2, 1944, and it was part of a report from Major General John Deane to General George Marshall back in Washington. The subject was cooperation with the Russians, all right, but the partnership was in the war against the Nazis, not in outer space.

Fifty years later, NASA was relearning the same lessons about dealing with the Russians, but with some bizarre self-imposed handicaps thrown in. When U.S.-Russian space negotiations began in 1993–1994, the Russians made an odd statement to their new partners: The Americans were told that the Russians would "feel insulted" if they encountered any NASA team members who seemed "too familiar" with the Russian language, Russian culture, or Russia's space technology. The American who passed along the request explained: "The Russians told NASA that any such people would obviously have been spies, and their presence would be considered insulting to Russia."

Michael Evans may have been a casualty of this policy; in any case, both he and the entire *ISS* project were the losers. A mid-level NASA manager in the flight design division, he had been loaned to the Pentagon in the early 1990s to take part in a commercial project. This project involved legally acquiring and evaluating Russian space nuclear power technology that had come up for sale in Moscow after the collapse of Soviet-style secrecy. The Russian design was called TOPAZ, and it had been built to power radar satellites targeted at Western naval forces. Evans was designated program manager.

By applying both the experience of earlier negotiators and his own common sense, Evans succeeded in his mission. The hardware was acquired on schedule and under budget, and the Russian engineers provided thorough documentation. Studies performed by American experts at the Sandia Laboratories on Kirtland Air Force Base in New Mexico provided significant insights into the technology and the Soviet plans to exploit it. "The US has leapfrogged the costly development cycle while expending only tens of millions of dollars," Evans wrote.

In 1995, Evans published a formal 116-page report on the project. "TOPAZ International Program: Lessons Learned in Technology Cooperation with Russia" was printed by the sponsoring agency, the

Ballistic Missile Defense Organization at the Pentagon. "We hope this document will be beneficial to other government agencies engaging in technology cooperation with the [former Soviet Union]," Evans wrote in the foreword.

Among the lessons learned, Evans stressed, were the importance of establishing formal technology partnerships and the value of persistence and patience in the face of the legislative and regulatory requirements that hinder cooperation. He also stressed the importance of close personal relationships and the need for technically trained interpreters. At NASA, in contrast, translation contracts usually went to the ex-girlfriend of a top official, who would immediately try to hire the cheapest translators available.

"The Russian work ethic discourages individual decisions except at the senior management level," he noted. This could be ameliorated by another suggestion: "Provide bonuses for Russians who achieve a high level of accomplishment." Evans added, "[Our] Russian scientists and technologists loved the concept."

Evans returned to NASA early in 1995 after managing this highly successful program. He told me that his first stop was the space station office, where he described his experiences and asked what role he could play in the project. He was stunned to hear that they didn't need him. All the senior positions were already filled, and the people that held them could read his report, if it was needed. Dismayed, Evans returned to his old nonmanagement job and applied to law school. Because he saw no future for himself in the *ISS* program and therefore had nothing to risk, he was one of the few people I interviewed who allowed me to use his name.

"I offered my experience to NASA upon returning to NASA and they chose not to apply any of the lessons learned from TOPAZ," Evans told me recently, although he could not recall the specific individuals whom he had talked to. He did remember their attitude, however: "There was an arrogance within NASA about dealing with the Russians," he told me. "I doubt whether anyone actually ever read the report," he added.

Although there were several guest lecturers and generalized training programs on "Russian culture' at NASA, most of the space workers had to learn it all over again for themselves. Many of them were bright enough to reach the same conclusions that other Americans had already reached years, or even decades, earlier. But sometimes the learning curve was very expensive.

It hadn't taken Bonnie Dunbar long to figure out that she wasn't in Kansas, or even Houston, any more. As a pioneer American cosmonaut-trainee in Russia in 1994, she was one of the first NASA workers to encounter the culture gap. In October 1994, she spoke publicly about her observations to a meeting of space professionals in Houston (that's why I can use her name).

The NASA office at Star City is in a building called the Profilaktorium, she explained, referring to a building whose name NASA workers needed considerable practice to pronounce without giggling. It is a "very modern" structure that had been built especially for the *Apollo-Soyuz* program 20 years earlier. A three-story building with a lot of antennas on the roof, it contains both offices (for NASA, ESA and others) and guest quarters. The astronauts had apartments in the building (there were two) that was closer to the training compound.

At the Gagarin Center, Dunbar pointed out, "paper is not a means of communication." Desks did not have "In" and "Out" boxes. There were no maps or phone books. Important things were learned by word of mouth. And the copy machines were *still* controlled. The paper going into the machine was controlled, and all of its products were controlled. Just as in the old Soviet days, a designated operator took your originals and the blank paper, inspected them both, ran the copies, and then inspected the copies, all before returning them to the person who needed to make the copies. This had once been standard practice throughout the USSR, but in recent years, the paranoid ritual has been abandoned almost everywhere.

For the first three months of training, the astronauts' days were spent in four-hour Russian language classes, plus sports and gymnasium activities for about two hours. They also did yoga twice a week for the first two and a half months. It had been used as a relaxation technique ever since Indian guest cosmonauts had introduced it to the Russians, and the chief psychologist really loved it.

On the subject of space food, Dunbar repeated the story that some space-bound food had been stolen from the *Progress* supply ships, and that as a result the *Mir* crews had been forced to use old rejects—particularly "fish aspic" (fish and transparent jelly)—for nourishment. Veteran cosmonaut Aleksandr Serebrov told Dunbar how much he hated breakfast in space, staring day after day at his can of fish aspic—(or, for lunch, his can of "chicken and prune aspic"). The food quality problem was in large part due to the loss of suppliers that occurred when the USSR split up.

Dunbar visited Baykonur, the launch site, and found it to be in a "third world country." There was no running water in the crew quarters. In every building she entered, she had to wash her hands with chloroform because of the cholera outbreak in a nearby city.

Dunbar had no patience with other Russian practices, such as blaming NASA for causing Russian delays. In an internal e-mail in late 1994, she described learning that the promised Russian science gear wouldn't arrive at *Mir* in time for NASA's first mission. "US folks here are a little steamed since we don't know what US delay they are talking about," she wrote. "Much US hardware is here and is being held up in customs, and wasn't due until [next month] anyway." According to Dunbar, "RSA is not being forthcoming with the facts, possibly for domestic consumption. If NASA accepts blame, it will be very demoralizing to all Phase 1 folks." NASA sidestepped the confrontation by deciding that the delays were "nobody's fault," so naturally they never got fixed.

Dunbar found the Russian attitude toward women to be reminiscent of the American male chauvinism of the 1950s, only a lot worse. As an astronaut and an official NASA representative, however, she had some protection. In 1999, a new crisis at the "culture gap" highlighted not only the Russian male views of women but, more importantly, their views on the public disclosure of the truth. It also highlighted how profoundly the Western space agencies had adopted the Soviet-style suppression of unpleasant news and of the people who dared to bring it to the public's attention.

In one case, Dr. Judith Lapierre, a 32-year-old Canadian medical specialist, had encountered violence and sexual assault during a 110-day mock space mission in Moscow. Worse yet, she and her non-Russian colleagues were ordered to cover up the incidents so as not to embarrass the Russians. As a result, she told journalists early in 2000 that she believed that the Russian space program had a long way to go before it could reliably take part in international space missions.

A handful of Russians and international partners had been isolated in a small spaceship simulator, where their interactions were observed by psychologists on television. Lapierre had been one of the three international test subjects who entered the mock spaceship at Russia's Institute for Biomedical Problems in Moscow on December 3, 1999. The three foreigners and one Russian joined the four Russians who had already been inside the three-room complex since early summer.

The experiment, called Sphinx-99, was designed to observe group dynamics under both routine and emergency conditions. The project's name is an abbreviation for Simulation of Flight of International Crew on *Space Station*. The sponsoring institute was Russia's leading space-flight medicine center. The small modules to simulate long space missions had originally been built in the 1960s.

Lapierre holds an M.S. in nursing and a Ph.D. in health sciences. She had studied at the International Space University at Strasbourg and performed psychological research in Antarctica. When the Russian medical experiment was recruiting international test subjects, she signed up.

Although she was not formally an employee of the Canadian Space Agency, Lapierre had applied for and received a grant from the CSA to document her experiences in the Russian experiment. She also hoped that the work would enhance her chances of actually being selected as a Canadian astronaut for the *International Space Station* program.

But then she was twice forcibly French-kissed by the Russian team commander, soon after witnessing a ten-minute-long fight between two Russians that left blood spattered on the walls. She took photographs of the blood-spattered wall with her digital camera and e-mailed them home to Canada. Lapierre and her associates from Japan and Austria then appealed to their sponsoring agencies to discipline the offenders. To their astonishment, they were told that such behavior was normal for Russians and that they should either put up with it or leave. They were also told that Russian manners prohibited them from making a public complaint.

It wasn't until 10 days after the assault that Lapierre and her teammates were given locks for the tunnel between their module and the Russian module. The Japanese participant was so upset by the lack of prompt and energetic support from outside that he quit the experiment altogether. Lapierre stayed, but she relayed her experiences to her husband and father in Quebec via e-mail. They subsequently ignited a Canadian media campaign on the subject of the inappropriate incidents.

Lapierre, who is just 5 feet tall, insisted that the controversial tongue kisses were not merely "friendly celebrations" and that she had vigorously told the Russian to back off. She told me that the commander had explained his intentions to her. "We should try kissing, I haven't been smoking for six months," announced the still-unnamed Russian. "Then we can kiss after the mission and compare it. Let's do the experiment now."

She dismissed the notion that the Russian thought that his actions were normal and acceptable. "Why did he try to pull me out of sight of the camera?" she asked. An American astronaut found Lapierre's accusations and complaints entirely credible: "Nobody wants to offend the Russians," he e-mailed me, "even when they totally misbehave."

The international test subjects knew that the Sphinx-99 isolation experiment would be tough, Lapierre continued, but they encountered unexpected hardships even in the simplest areas. "We were promised hot water, but for two and a half months we had to shower in cold water," she said. "We had to eat kasha [wheat gruel] every night, there was no variety or international flavor to the food." She sent home digital photographs of what she called "the cockroach invasion," as well as the subsequent "head lice invasion."

More critical to the tensions that arose were communications problems. "We were promised English as the official language of the mission, but Mission Control didn't speak any English," she said. She was allowed e-mail and brief weekly phone calls to her family.

When the story broke in Canada in March 2000, I telephoned Lapierre in Moscow as part of an effort to write her story for a U.S. audience. Her team had left the modules on March 22, and we talked just a few days later. Over the following months, we became friends. She found that her public protests left her with few friends in Russia or in Western space agencies.

"The problem was not so much the incident itself, but their reaction to our concerns," Lapierre told me from her hotel in Moscow. "If people are to work together, they have to share some basic principles and rules."

Lapierre told me that although some press accounts were distorted and exaggerated, she was grateful for the media coverage. "If it had not been for the press, it would have been minimized, and nothing would have been done," she said. In the end, the whole experiment was a flop: "This was a chaotic field study, not a scientific experiment," Lapierre told me. "They were not ready to host an international study."

Once the experiment ended, Lapierre was eager to talk to the news media about what really happened, and she now fears that her words have scuttled her chances of ever joining the space program. She described her research in the Russian modules to a Canadian television program: "I was doing a study on human interactions and adaptations to life in confinement," she said. Its purpose was "for selection of astronauts in the future, looking at what factors are essential."

From a preliminary evaluation of her data, she announced one conclusion. "Sharing some basic values and principles is the major issue if you are to work in an international space station," she said. "There should be strict rules and codes of conduct, because simple rules of society should apply in any confinement or space experiment."

In an e-mail sent to their sponsoring agencies, the foreign members of the crew had complained about the lack of an official response to the Russians' misbehavior. "We had never expected such events to take place in a highly controlled scientific experiment where individuals go through a multistep selection process," they wrote in a letter published by the *Globe and Mail* in Toronto. "If we had known ... we would not have joined it as subjects."

In December 1999, when Lapierre's team first entered the modules, Valery Gushin, the scientific coordinator of the project, clearly exhibited some cultural attitudes about women in Russia that in hindsight could have been seen as warnings about the problem. "Men, they have some expectations from women," he told a Canadian television team. "They want them to be more like women, not just partners. At least Russians [do]."

Following the incident, Gushin blamed Lapierre. She told me that his official report stated that she had "ruined the mission, the atmosphere, by refusing to be kissed." She should have been taken out, he wrote, and he also insisted that the foreigners had caused the fight.

After the first press attention to Dr. Lapierre's complaints, Gushin told reporters that it had all been a cultural misunderstanding. "In our culture, there is no difference between kissing on the cheeks and kissing on the lips," he explained. "Especially if it's connected to celebrating a holiday or a party, then this is nothing.

"This perhaps shows that there must be special training for men so they can understand the various taboos of women from other cultures," Gushin admitted. "Our men probably have to learn to be more prudent and understanding."

The Russians also called the men's clashes "friendly fights." But the non-Russian crew members were so alarmed that they hid all the knives in the modules in case the fight resumed.

Gushin criticized Lapierre for cultural insensitivity in making her complaint public: "In our culture, it is taboo to wash your dirty clothes in public." Lapierre described the attempted cover-up. She told Canadian reporters that a preliminary summary of the mission had whitewashed the incidents, and that she had been required to sign it before the modules would be opened to let them out.

Russia's INTERFAX news agency presented an account of the incidents that varied significantly from that of the foreign test subjects. It stated that Lapierre "considered a Russian male colleague's attempt to kiss her on New Year's Eve as sexual harassment." Interfax then quoted Gushin explaining that the problems had occurred "while the crews were adjusting to isolation." According to Gushin, "They displayed excessive irritability, briefly lost self control, were overly emotional, and too categorical in assessments," he said.

The Russians who had already been in the modules for six months were the ones who had misbehaved, however. Gushin, it seemed to Lapierre, was blaming their behavior on the newly arrived foreigners. She also complained to me about other Moscow press coverage of her complaints. "They are only protecting the Russians, not trying to understand our feelings," she said.

After the end of the experiment, the Russian crew felt sad, perplexed, and confused at the fuss over what they considered to be such a "small thing," said Dr. Raye Kass, a professor of applied human science at Concordia University, in Montreal, Canada. Kass had been one of the "principal investigators" for the project, and there weren't many others. NASA had been invited to participate in the experiment, but after reviewing the proposed protocols, it had declined. So had research teams in France and Ukraine.

"We didn't believe in the past why NASA didn't trust the Russian medical data," Lapierre told me. "Now we know."

An American astronaut and medical doctor privately explained to me what Lapierre meant about NASA's view of the Russian medical data from space flights. "It was widely known that many of the cosmonauts didn't take the medical experiments seriously and enjoyed messing with the data," the astronaut told me. This behavior involved "taking another's place to confuse the data, cooking data, [or] refusing to participate."

"I cannot remember any astronauts doing the kind of malicious mischief routinely practiced by the cosmonauts," he concluded. Author Bryan Burrough (*Dragonfly*) also reported that both Russian and American medical specialists consider Russian long-duration space-flight biomedical monitoring to be "total junk." According to Burrough, when an American doctor asked a Russian colleague for the medical database that the Russians use for long missions, he was told that they use the NASA *Skylab* data from 1973–1974.

Lapierre expressed disappointment over the isolation that she suffered. "I expected to be in better hands," she told the Discovery-Canada science program, referring to the Russian reputation for psychological support during space missions. "But I'm doubting today what kinds of psychological support they are giving to cosmonauts, if they are giving any, because I didn't get any from them."

Non-Russian women's experiences with Russians on space missions have so far been a lot smoother than Lapierre's experience. In 1996, an American woman scientist, Dr. Shannon Lucid, spent six months aboard *Mir* with two Russian cosmonauts. Their relationships were by all accounts cordial and cooperative. An English woman and a French woman have also visited Russian stations. The team assigned to become the fifth crew aboard the *International Space Station* also consists of one American woman, Dr. Peggy Whitson, and two male Russian cosmonauts. They are expected to fly into space in early 2002.

Actual space crews have very different psychological preparations for their missions, making a repetition of Lapierre's experiences unlikely aboard *ISS*. The crew members train together for a year or more before the mission, and they are in much closer contact with specialists during the flight.

Lapierre stressed to me that she was criticizing the program management, not Russian cultural features. "It wasn't a Russian culture thing," she explained, "it was a personality trait and crew selection failure." When the Japanese researcher walked out in January, nobody replaced him. The following month, the long-term Russian crew left, except for a life-support systems engineer. "He came over and stayed with us," Lapierre said. "We got along fine."

Full-time NASA employees rarely made Lapierre's mistake of telling inconvenient truths in public. When an American was badly beaten by Russian plainclothes policemen in 1999, reportedly suffering a mild concussion and several broken fingers, NASA officials claimed that they filed official protests. When I investigated, I was told that the protests were "off the record," so they couldn't show me the copies. My requests to interview the victim were refused. As recently as December 2000, the NASA office in Moscow was issuing urgent warnings about harassment by Russian police. It advised workers to always carry emergency contact information with them, and to travel in groups. It also advised, "Do not get into a car with Russian [police] if you can help it." As a reassurance, it added, "The Embassy has a roving patrol who the Marines will send immediately to assist you if you are having problems."

In the most extreme cases, NASA workers realized, telling the truth in public could be physically dangerous. Todd Breed was an enthusiastic young NASA space engineer assigned as a liaison to the Russian space industry in 1994. He sent back a series of insightful and candid evaluations of the situation. "There sure are a lot of new cars in front of the Russian Space Agency building," he told me during one of his infrequent Houston visits, "and all the Russians have expensive new suits."

One memo, dated November 22, 1996, slipped out onto the Internet and was widely read. "The financial problems with Energia are bigger than we thought," Breed had written. He was quite correct, but NASA apparently didn't want to know this. According to the memo, RSC-Energia "owes close to $100 million to its subcontractors for past work on other programs, and most of them will not accept any new work until the debts have been paid." The subcontractors faced imminent bankruptcy without the funds, Breed continued, and added: "Only the 25- to 30-year history that many of them have of working with Energia has prevented some of them from shutting down altogether."

The Russians were furious that these assessments had reached the public, and they banned Breed from visiting their facilities. NASA, also piqued, would not support Breed and declined to demand that the Russians rescind their ban. Expulsion from a project that he loved, and also some personal problems in Moscow, sent Breed into depression. He went home on a medical leave, totally ignored by NASA. A few months later, he was found dead in a Vienna hotel, an apparent suicide.

These cultural-interface experiences cover the full spectrum from rewarding and thrilling to dismaying and frightening. But a common theme in all of them is that the Russians are different.

"The Japanese, Americans, Canadians, and we in Europe have approximately the same technical culture," noted Jorg Feutel-Buechl, the European Space Agency manager for human space flight. " But the Russians have a completely different approach to these things."

Mainly, he observed, Russians don't leave a "paper trail," the extensive documentation process used in the West for quality control and post-accident investigations. "They made changes and adaptations without extensive paperwork," he explained, describing how they modified a German-made computer for the *Service Module*. "I wouldn't say they have not been careful enough. But they have been rather courageous in their approach."

NASA encountered some of the consequences of this "courageous" approach when it asked a Russian team to describe the design principles of a space rendezvous abort program on the *Soyuz* computer. After several months of waiting, NASA raised the subject at a general meeting and got a stunning answer: "The software was designed a long time ago and no one really knows how it works." A NASA engineer recorded the results of his own inquiries in a memo: "It seems that no two people gave the same answer about the software or its origin."

"Their data is all handwritten in little green books in the pockets of operators in Moscow," NASA official James van Laak explained to Congress in 1997. "They have no published summaries of events."

An American scientist who worked with a Russian team during shuttle-*Mir* corroborated these assessments. "They had all these procedural weaknesses," he told me. "Half the time they didn't have any procedures, and then when they did they often didn't follow them. And they seem to be very much stuck at the point of 'it works, this is what we've been doing for years, there's no need to change anything, we'll just brute-force our way through it like we always do.'

"Their flight crews are fearless almost to a fault," he continued. "And they all seemed to have a very dry, almost British sense of humor."

Cosmonaut attitudes toward their "newbie" junior partners in space were often patronizing. Michael Foale recalled always being "mother-henned" by his crewmates. "They think Americans and Westerners generally are soft," he told a BBC program. "They believe that Russians have a natural ability to suffer, to take hardship and surmount it. They think, 'oh, we have to make it easy for that person, this person's going to be unhappy and miserable if it's not easier for them than for us.' It's a feeling of condescension and patronizing, and Sasha—I mean, I love him—but he—I always had to laugh—he would always try to shelter me from anything that was going on on *Mir*. And the institution is trying to shelter the foreigner from anything that's going on." That especially meant shielding foreigners from 'bad news' that might worry them, an attitude we've encountered again and again on these pages.

Sometimes Americans went to extreme lengths to immerse themselves in Russian culture, to learn the language, and to live away from the "NASA compound" of fellow countrymen. One such engineer was dismayed by what he encountered. "I have realized that no one became successful or gained a position of authority under the USSR by being honest, straightforward, open, hardworking, and being nice," he told me. "You had to learn to lie, backstab, cheat, steal, while creating an air of confidence and competence. I am convinced that most Russians do not even know when they are lying, it comes so easily and naturally. Countless daily experiences. Many Russians have been described to me as having a split personality. One [personality] that lives the lie of the society and another that you *may* be able to hold on to internally or with your wife. These are the 'successful' people, still in charge of the whole country. In business, government, Mafia, indistinguishable—and even in places like the TsUP, etc."

Others saw this aspect of Russian culture in a more favorable light. "Nowhere, do I find, is there such an enormous gap between the facts and the truth as in Russia," wrote Steven Jones, author of *Manual for Negotiating with Russians*. He explained: "After centuries of despotism, arbitrariness, and abusive government, nothing important or true or even potentially sensitive in any way has ever been written down or shared. The only way to find out what is really going on over there—'the truth'— is to become close enough and trusted enough to real Russians that they will feel safe enough and be motivated to explain what's going on to you."

Jones's company, East-West Business Strategies, runs seminars for NASA workers going to Russia. Russian culture is incredibly complex, he explained to me. The orientation program for Japan "takes a day and a half, and when it's over, the participants report they are actually able to apply what they've learned with huge effect." In contrast, "I've been doing Russia in three days, and we've known for several years now that we need at least four to five days just to cover the factual and theoretical material, let alone transfer skills and confidence in any meaningful measure." Even though the seminar's student evaluation forms regularly proclaim that "we need more of this" and ask "why isn't my boss here?" NASA policy has been that the higher officials are too busy to take the time to learn this material.

Space engineers, accustomed to immense physical distances between satellites, Earth, and other worlds, must now struggle with the immeasurably greater separation between human cultures. Most have been bright enough and dedicated enough to eventually overcome these barriers. But again, NASA's initial attitude, claiming that it already knew all that it needed to know, made the entire process more time consuming, more expensive, and more dangerous.

14

Staking Out
the Orbit

*"It takes two to speak the truth—one to speak, and
another to hear."*

Henry David Thoreau

*"Real knowledge is to know the extent of one's
ignorance."*

Confucius

Nikolay Ganzen and I had a lot in common. We both liked French
wine. We were both big men, and we both had a big sense of
humor. But most of all, we both had unearthly minds. One way that this
showed up was in the things that we, almost alone among our compa-
triots, thought were humorous. We were accustomed to seeing hilari-
ous irony in situations that were evidently not amusing to others.

He spoke Russian and I was raised with English, but we shared a
common language that few people in either of our countries understood; in Russian, it was called *ballistics*, in English, *orbital mechanics*.
Based on mathematics, this specialty defines the curving paths that
objects follow as they swoop through space under gravitational forces,
steered by periodic rocket firings.

It was in this unusual language that the final designs for the foundation of the *International Space Station* were expressed.

By early 1996, the Russian-American space marriage had been con-summated with the first shuttle dockings to *Mir*. Hardware for a dozen different modules was taking shape. The design and assembly sequence of the *International Space Station* had been defined. Now the questions concentrated on how to carry out these plans.

Ganzen and I first met at a quarterly Technical Interchange Meeting, or TIM, in Houston. Two dozen different teams of specialists from both countries met in row after row of cubicles to thrash out spe-cific topics. There were questions about communications, life support, logistics, "who was in charge of what," flight crew activity scheduling, and so many more topics that often a particular panel would splinter into subgroups assigned to more highly specialized questions.

Our panel was flight design, and we were supposed to specify the exact orbit into which the very first *ISS* module would be launched. The flight, which was planned for the following year, didn't actually occur until November 1998. By then, as it turned out, I had already left the program to pursue a full-time career in writing and consulting. But NASA ultimately followed the plan that we came up with at the meeting.

My job in 1996–1997 was to coordinate the orbital design of the shuttle mission that was to carry up the first U.S. segment a few weeks after the Russians launched the first part. To do this, we had to deter-mine what kind of orbit we needed to end up in.

Orbits have been my professional focus for most of my life. In col-lege, I studied the mathematics of orbital rendezvous. That was during the period of time in which the *Gemini* spacecraft actually performed the maneuver in orbit. In graduate school, funded by NASA, I studied computational techniques for planetary swing-by trajectories. In the post-*Apollo* cutbacks, NASA wasn't hiring, so in order to continue to do space-flight-related work, I joined the Air Force, where I kept tabs on Soviet spy satellites. I issued alerts to my base about when the satellites would be overhead, so that we shouldn't be engaging in sensitive activi-ties outside. When Soviet space stations maneuvered in orbit to line up for new launchings from the ground, I compared the maneuvers to their standard patterns in order to forecast the launchings themselves. Accurate forecasts of Soviet launches could be thus made in the absence of any official announcements.

Orbits remained the focus of my professional life, and of some of my hobbies. At Mission Control in 1982–1983, I led a contractor team that supported the first shuttle experiments with free-flying payloads and with rendezvous. I then literally wrote the book that explained astronaut procedures for space shuttle rendezvous. When the Long Duration Exposure Facility (LDEF) satellite was retrieved in 1990, I was the lead guidance and procedures officer at Mission Control. Several DoD shuttle missions also needed my specialty, but that's all that I can say about that. I also wrote the console reference handbook and the training plan. In the early 1990s, I tracked down, collected, and annotated several dozen key reports for a comprehensive "History of Orbital Rendezvous" for the library at NASA. It's safe to say that I often dreamed about space orbits.

Describing space orbits verbally, without using gestures, has always been a challenge. It's sort of like trying to get someone to describe a spiral staircase while sitting on his or her hands. But with the right analogies, it can be very illuminating. Since I've done it for my mother, the Boy Scouts, visiting NASA managers, miscellaneous politicians, and even an astronaut or two, I know it can be done.

Think of a space orbit as a wire circling a globe. One end is tilted up from the globe's equator (and so the other end is tilted down). But the center of the wire circle has to match the center of the globe. The circle defines a flat plane in which the satellite moves, like a bead along the wire.

The object of our attention, the vehicle that we were designing an orbit for, was called the *FGB*, or *funktsionalno-gruzovoy blok*. This means "Functional Cargo Block," and it was the Russian designation for the 20-ton vehicle that NASA at first just called the "*Salyut* Tug."

The *FGB* may have been new to NASA, but these vehicles had been a part of the Soviet space program since the mid-1970s. Their ancestor was called the *TKS*, and it was designed to ferry both supplies and, with the help of a capsule mounted on the front, cosmonauts to military space stations.

My friend Ganzen had been a part of this, as a team member of what is today called the Khrunichev Center in the Fili neighborhood of Moscow.

Back when the Russians first joined the *ISS* partnership and the idea of using an *FGB* came up, NASA soon realized that the Russian government wasn't going to fund it; the United States would have to. Besides, since for domestic political reasons an "American" module had to be the cornerstone of the *ISS*, NASA would have to hold the deed of ownership for the *FGB*. For the sake of that piece of paper, NASA agreed to send the Khrunichev firm more than $200 million via NASA's space station contractor, Boeing.

So officially the *FGB* was an American module, even though it was built in Russia according to Russian designs. The design included an Earth-to-space command and control system, through which Mission Control operated the vehicle in orbit. But the Russians never told NASA what the command codes were, even after they were directly asked to do so. Such behavior might indicate whom *they* considered the *FGB*'s real owner to be. And NASA's own internal documents seemed to agree, lumping the *FGB* into the "Russian segment" of the station.

I annoyed a whole series of NASA managers and publicity officials by innocently asking whether there really shouldn't be an American flag on the *FGB*, since it was after all an American spacecraft. The question was considered tasteless, ridiculous, and meaningless, probably because nobody knew the answer. I finally was told that there was in fact an American flag, about four inches high, along the *FGB*'s bottom rim, and I later found a handful of photographs that showed it. The Khrunichev factory had charged the United States an astronomical amount of money (one official told me it was about a million dollars) to paint it on.

In the original "Russian Alpha" design for the space station, which dated from 1993, when a combined U.S.-Russian space station was first discussed, the *FGB* was to serve as the stabilization and control module for the early space station structure in what was called the "man-tended mode" (the early flight phase, when crews were present only part of the time). That was before the arrival of the life support equipment that would permit a permanent human presence, although some power modules and a laboratory module were already on board. A competing spacecraft, the Pentagon's so-called *Bus-1*, promised similar capabilities, but at the cost of almost a billion dollars, so choosing the Russian option made economic sense.

Over the next three years, the design of the station evolved, and the purpose of the *FGB* underwent a major but unheralded transformation. The arrival of Russia's *Service Module*, with its control and life support functions, was moved so that it occurred much earlier in the assembly sequence: four months after the launch of the *FGB*. By that time, only a single module—the U.S. *Node-1* connection block—would be hooked to the *FGB*.

The reason for pushing up the date was based on the need to mate the two Russian modules—(the *FGB* and the *Service Module*) in space. The two modules were launched separately aboard powerful *Proton* rockets. But only *FGB* had the steering rockets and guidance sensors to chase down and dock with another vehicle. The *Service Module* could "cooperate" by holding its position and sending out answering radar pulses, but it couldn't maneuver at all.

So for the *FGB* to be able to fly its docking profile, it couldn't be encumbered with too much extra baggage. The *Node-1* module could be mounted on its back end, and the *FGB* would still be able to chase and capture the *Service Module*. But adding anything more would make the *FGB* too heavy and its weight too off-center for it to steer properly.

This reasonable alteration of the assembly sequence led to one very embarrassing question, however: If the *Service Module*, with its control computers and thrusters, would be sent into space only a few months after the *FGB*, what purpose did the *FGB* serve that was worth its $200 million price tag? Why not just launch the *Service Module* alone, send up a shuttle with the *Node-1*, and proceed with the assembly from there?

NASA headquarters vetoed that suggestion. The *FGB* would remain part of the sequence. No other options would be considered.

It turned out that the decision to keep the *FGB* was pure public relations, or, more specifically, congressional relations. This went to the core of the concept of the *International Space Station*, which was designed to be led by the Americans. If Russia's *Service Module* was the first element placed in orbit, the Russians could proclaim that the new station was a Russian station, with subsequent American add-ons. Congressional and public support could evaporate.

So the supreme requirement for the *ISS* design was a good political appearance. Whatever the reality behind the way the station was being assembled, it had to *look* as if it was mainly an American system, with Russian add-ons. Even if that wasn't true (and in the beginning it wasn't), it had to be made to look that way so that Congress and the American public would perceive the desired image, not the reality.

So an American module just *had* to go up first, or at least ahead of any purely Russian module. The *FGB* launch was an acceptable mongrel, "half Russian and half American." And the Russians did pay for the launch, which was only fair considering how much the United States overpaid for the module itself. The follow-on all-American *Node-1* then established that the United States' contribution was the station's cornerstone, even though once *Node-1* was in orbit, it didn't do anything useful until the *Service Module (SM)* arrived. The *Node-1* really didn't accomplish anything significant until the U.S. *Laboratory Module* showed up a year after that.

The plans were that when the *SM* arrived, the *FGB* would be demoted to a storage shed and spare fuel tank. Its electronic systems would be deactivated, their functions assumed by the *SM*. One consequence of this—which would later come back to haunt NASA—was that unlike the *SM* and other mainstream *ISS* modules, which were designed for a full 15-year lifetime, the *FGB* was considered only temporary, and so its equipment didn't require anywhere near that longevity.

But the only thing Nikolay, myself, and our fellow ballisticians had to worry about was the orbital consequences of these designs. These questions of policy were way above our heads, or, as we used to say, "way above our pay grade." We were faced with a more immediate technical question when selecting the initial orbit for the *FGB/Node-1* complex: Where would it lie in relation to the orbit of *Mir*?

Mir was occupied by teams of cosmonauts during our design process that year, and during the next year as well. And although the Russians kept assuring NASA that they would close down and deorbit the aging station once the assembly of *ISS* began, we knew that there would be at least a few months of overlap.

Nikolay and I had essentially two options. We could place the *FGB/Node-1* complex in an orbit near that of *Mir*, which could make station-to-station transfers possible, or we could place it in an orbit as far *away* from *Mir* as possible. In evaluating these possibilities, we were assisted by a senior ballistics expert from RSC-Energia named Ludmilla Chaikina. A diminutive, slightly graying woman with a shy smile, Chaikina had a classic steel trap mind and an iron backbone when it came to her own views. We all got along great.

As I discussed these strategies with my management, I made sure that everyone understood the implications of each choice. The Russians had already tried to delay the delivery of the *Service Module* and to substitute the existing *Mir* as the "foundation" on which the new *ISS* modules would be assembled. NASA managers rejected this proposal immediately, since the impression that the *ISS* was an "add-on" to the aging Russian station would have been political suicide. But if we placed the *FGB/Node-1* complex in an orbit close to *Mir*, we would be giving the Russians the option of creating a fait accompli. If they announced that the *Service Module* had suddenly been "unexpectedly delayed," they would be able to fly the *FGB/Node-1* complex over to *Mir* and dock it there, since the orbit we had chosen permitted such a maneuver.

The joint meeting began with the aim of listing the operational requirements that would determine the selection of the best orbit. We were expecting the Russian side to propose that the chosen orbit should allow station-to-station transfers, thus enabling them to move two of the science modules from *Mir* to the *ISS* for extended operations. The Russians had also been talking about the need to retrieve and transfer up to five tons of other miscellaneous equipment—research equipment, cameras, supplies, even musical instruments and memorabilia—from *Mir* to the *ISS*.

Ganzen wasn't interested in these options. To our total surprise (and delight), he carefully described a number of technical reasons why he, too, wanted the *FGB/Node-1* complex to be as far away from *Mir* as possible. Chaikina agreed with him.

To understand how orbits can be close to or distant from to each other, you have to abandon Earthside notions that distance is a measurement of physical separation. In space orbits, everything is moving very fast and distances are changing constantly. What "distance" really means to a ballistics nut like Nikolay, Ludmilla, or myself is the amount of effort needed to get from one orbit to another. That effort is usually measured in terms of "change in speed" or "delta velocity," or just "delta-V." In practice, that means burning a portion of your on-board rocket propellant supply, which is always very limited.

Both *Mir* and the *ISS* were going to be circling Earth about 200 miles above the atmosphere. These circles were "flat," in a single plane.

But that plane was tilted relative to Earth's equator, about halfway (by 51.6 degrees, to be precise, but you can forget that exact number).

If both *Mir* and the *ISS* were in the same plane, the transfer of goods between them would be relatively easy. Raising or lowering an orbit is simply a matter of speeding up or slowing down. You might need to vary your speed by about 10 or 20 feet per second, relative to your normal orbital speed of 25,000 feet per second.

The Russians had performed exactly such a trick in 1986, when they had just launched *Mir* and were at the very zenith of their space-flight power. The first pair of cosmonauts who boarded *Mir* later undocked their *Soyuz* and flew over to the *Salyut*-7 space station, where they stayed for about a month. After performing a few more space walks and loading their *Soyuz* with half a ton of specialized equipment and memorabilia, they flew back to *Mir*.

This wasn't as simple as it sounds. The two stations had to be kept in a relatively tight formation, so that their orbital planes didn't shift apart. When the time came to initiate the transfer, the station holding the crew was put through a very fuel-expensive firing of its rockets to push it toward the other station. The crew then detached in their *Soyuz*, and a day later, they used the fuel in the *Soyuz* to make the final rendezvous with their destination. As they were docking, the station they had just left went flying past, a few miles beneath them.

But for the *ISS* and *Mir* in the 1997–1998 time frame, station-to-station transfer wouldn't be so simple. That's because of another variable of the orbit, the "hinge line," along which the orbital "tilt" is measured against the equator. That line can be in any orientation as long as it passes through the center of the Earth and two points on the equator on opposite sides of the Earth. Both orbits can have exactly the same tilt—the identical inclination angle to the equator—but if the hinge lines at which the orbits are tilted are too far apart, the orbits will be hopelessly far apart when it comes to trying to go from one to the other.

Prove it to yourself with a simple demonstration. Hold a CD-ROM (or a plastic plate) in both hands, with your fingers at the 9 o'clock and 3 o'clock positions. Tilt the end closer to you down about halfway (the end farther away from you tilts up), and remember the pathway along the disk's rim. Now, again from a level position, put your fingers on the 12 o'clock and 6 o'clock positions, and tilt the left edge upward about halfway (the right edge goes down). Look at the pathway along the rim and compare it to the previous case. They both are tilted up about the same amount, but the two pathways—the orbits, in my analogy—cross each other at steep angles. This is because their hinge lines are different. And a spaceship doesn't have enough rocket power to jump from one orbit to another.

Before my colleagues in the ballistics brotherhood make too much fun of me, I should point out that the angle of tilt of the CD-ROM is equivalent to the orbital inclination of a space vehicle, and that the orientation of the hinge line is technically called the "longitude of the ascending node" or "right ascension." The first time Nikolay, Ludmilla, and I had to get the NASA interpreter to figure out those words, we watched her nearly choke. So we just made a few sketches and invented our own "Runglish" ballistics terms.

Although *Mir* and the *ISS* were each following a simple near-circular orbit in space, the paths that they each traced out on Earth's surface were complicated by another type of motion, that of Earth's own rotation. And locations on Earth's surface, particularly Russian space tracking sites, would also have a crucial influence on the selection of the orbits.

If Earth didn't turn on its axis, the "ground track" of a satellite would just be a big circle around the globe. It would be tilted to the north and south of the equator at the same angle as the orbit was inclined at its hinge line. When plotted on a flat map—say, a Mercator-type projection—this track would look like a sine wave, first curving to the north, then turning toward the south, then curving northward again.

Now add in Earth's motion as the circling satellite continues to trace a line on Earth's surface. It takes about 90 minutes for the satellite to complete one full circuit. When it gets back to its starting point, Earth's surface has moved eastward. It moves 360 degrees in 24 hours, or 15 degrees per hour. So after each satellite circuit of 1.5 hours, the Earth's surface has moved about 1.5 times 15 degrees, or 22.5 degrees, farther to the east. OK, fellow ballisticians, I know it's actually 360 degrees in 23 hours 56 minutes (Earth's "day" measured against the stars), but we're trying to communicate concepts here, not details!

The resulting ground track, over a period of a day or so, looks like a squashed Slinky, or maybe like a "ripple tank" physics demonstration from high school. Repeated copies of the sine wave pattern appear, each displaced slightly from the previous one. Now take a deep breath.

However complex the final pattern, it came from only two simple motions: the satellite moving straight forward in its inclined circular orbit, and Earth going about its business of making sunrise and sunset shows for the observers on its surface.

There was one more motion that needed to be considered over long periods of time, say weeks or months. For various subtle reasons of geophysics ("Earth's equatorial bulge" is the theme here), the orbit of a satellite near Earth is slowly twisted in space. For objects such as *Mir* or the *FGB*, the hinge line, or the orientation of the orbital plane in which they travel, was moving westward about 6 degrees per day. For any two satellites in similar orbits, the motions would nearly match each other, so the orbits of the two satellites would stay at about the same relative position. But for observers on Earth's surface—at a radio tracking site or a launch base, for example—that motion means that on average the satellite will repeat its passes about 24 minutes earlier each day (Earth rotates 15 degrees in one hour, or four degrees per minute). So launching into the proper relationship with *Mir*'s orbit was shooting at a target that was moving in several different directions at different rates. Nobody ever said it was easy.

What Ganzen wanted for the *FGB* module was a ground track that was always west of *Mir*'s track. If, for example, *Mir*'s track carried it northeast across Africa, then across Siberia, and then southeast across the Pacific, the *FGB*'s track during the same period would run from the eastern Pacific northeast across the Atlantic, across Europe, then southeast across the Middle East and on to the Indian Ocean. And every hour and a half, these paths would be repeated, only Earth would have rotated farther east, so that the tracks on the ground would have been displaced to the west.

Viewed on a map, Ganzen's logic was very clear, and he explained it to us in detail. *Mir* would cross near a tracking site in Russia and be in radio communications with that site for 5 or 10 minutes. An hour and a half later, it might again cross Russia and pass within range of the tracking site. Sometimes a tracking site might catch sight of *Mir* four or five times in a row before Earth's rotation carried the site too far to the east to be close enough to *Mir*'s next ground track.

But then the same tracking site would get its first visit of the day from the *FGB*. There would be a few more passes before the *FGB*, too, had shifted too far to the east to be within range.

"At any ground site," Ganzen explained through an interpreter and through sketches on paper, "all the *Mir* passes for each day must be completed before the first pass of the FGB occurs." He explained the practical reason for this: "We use the same radio equipment for both vehicles but it takes about an hour to manually reconfigure the receivers."

According to his calculations, Ganzen told us that this convenient situation would occur if the two orbits were shifted to be about 140 degrees apart. The *FGB*'s orbit would be "west," as measured along Earth's equator, of *Mir*'s.

Since the shuttle would blast off into this same orbital path through space just two weeks after the *FGB*, we were under some constraints of our own. There was the desire to have the landing take place during daylight (it would also be "nice" to have the launch in daylight, but that wasn't nearly so necessary). Our rendezvous sequence required making certain course corrections based on visual sightings of the target, and since these occurred at specific day/night points in each orbit, our ultimate arrival next to the *FGB* was locked into "orbital daylight." The Russians required that the actual contact occur over Russian radio sites, so that they could send up backup commands as needed.

Requirements had to be balanced, then carefully prioritized, and adjustments had to be made in the normal sequences of both sides' schedules. But over a period of six months, Nikolay, Ludmilla, myself, and the rest of our team developed a workable compromise that satisfied all of the important requirements. We then began briefing our management concerning the results. They liked it, and I received a "sustained superior performance" award for the process. Best of all, for us on the U.S. side, it was the Russians' own requirements that put the new station so far "out of plane" with *Mir* that any transfers from *Mir* to the *ISS* would be impossible.

One thing that bothered me was that even though all of us "worker bees" thought that the orbital plan was hot stuff, it would still be easy for top management to undo it on a whim. As I kept circulating reports throughout the U.S. side of the space station project, warning that we needed formal program approval to "lock down" the plan, I kept on getting the same old looks. I was being obsessively paranoid again.

A year after I'd left the program, with the launching only two weeks away, somebody high in the Russian space hierarchy suddenly realized that the agreed-upon orbit made cross-transfers impossible. On November 4, 1998, I received a short e-mail from a friend: "[Energia] has complained that they were not involved in the *FGB* time of launch development and suggest [it] be moved 11h40m." Other messages said that the proposal had already been rejected by NASA.

Of course, Energia had been involved in the original discussions, in the person of Ludmilla Chaikina. No, if Energia wanted to change the orbit, it was because it realized that it couldn't afford to build any new add-on modules. It wanted to use old ones that were already hooked to *Mir* to fulfill its promises to *ISS*.

An even more alarming possibility was that the move was an opening gambit in a game that would see the *SM* delayed still further, then canceled. The old *Mir* would then be the "obvious" replacement facility, and we'd be back at the December 1995 proposal that was so vigorously rejected by NASA. This interpretation was supported by the timing: only two weeks before launch, and right in the middle of NASA's exuberance over the spectacular public relations triumph of the John Glenn space shuttle mission, then in orbit.

Although everybody assumed that Russia's new proposal would be rejected, by the beginning of the following week, a worrisome rumor began to spread that the demand would be settled "at the White House level." That meant that politics, not practicality, would rule once again.

NASA officials said that the proposal was a modest one. "The Russians said they wanted to transfer some equipment off of *Mir* to the *International Space Station*," Dwayne Brown told the *New York Times*. "The request only concerned moving internal equipment from *Mir*, not transferring modules from *Mir* to the new station." But the same article quoted me as saying that once the "modest" intention was satisfied, far more profound and worrisome station-to-station transfers were possible. For example, environmental experts at NASA were convinced that the chemical contamination inside the *Mir* modules far exceeded the permissible levels for the *ISS*.

One NASA worker showed that he'd been learning over the years. "This Russian tendency toward changing things at the last minute reminds me of the book, *You Can Negotiate Anything*," he e-mailed me. "One of the key bits of advice in that book was to slip in seemingly small changes at the last minute."

The White House rumor turned out to be false, and NASA officials rejected the Russian proposal. After a "frank and candid" discussion, program manager Randy Brinkley told news reporters on November 13 that "the conclusion was mutual" and that "both sides concluded that it did not make sense."

But the Russians pulled one last trick on unsuspecting NASA officials. Since the FGB still didn't have a name, Russian officials suggested that it be called *Zarya*, which means "dawn." They volunteered no information on the name's origin deep within the history of their space station program—it had been the original name of the USSR's first space station, in 1971. NASA had no idea of the historical significance of the name, so it agreed. By agreeing, the agency tacitly confirmed that the *FGB* was and always had been a Russian module, and that the tiny U.S. flag painted on the side was the diplomatic equivalent of a fig leaf.

During the long design process, one vivid image had stuck in the back of my mind. Given a launch date, we were compelled to pick a precise launch time because of the desired relationship with *Mir*'s orbital plane. All subsequent launchings throughout the history of the program would have to align themselves with the orbital plane that we had specified.

Hundreds of teams of engineers in both countries had spent years developing hardware items, step-by-step procedures, complex control software, and other palpable contributions to the huge space complex. But our contribution was the essence of invisibility, an absolutely weightless sine qua non. Its importance was paramount, I told myself, because every other item being built by every other worker would be steered by our numbers. It didn't sound very objective, but maybe because we had spent so much time with cold-blooded calculations, we felt like basking in hot-blooded self-congratulation to restore some balance.

"Think of us as advance scouts for a huge army," I told Ganzen and Chaikina one afternoon when the day's negotiations were completed. "Behind us are miles-long columns of personnel and equipment. Ahead of us is a vast empty meadow, where we will soon make camp."

There we were, walking in our imaginations onto the trackless field where a great assembly of tents and trenches would soon fall into place. But where would the place be located? Aligned with what base marker?

"We pounded in the first stake," I bragged, going through the motions of wielding a hammer while holding a wooden pole. "This point right here will be the origin, the Square One, the cornerstone for all future construction which will depend on this baseline." We all smiled. It had been an interesting problem in ballistics, and we were very proud of our solution.

Late the following year, 1998, I resorted to another ballistic analogy when I testified at a congressional hearing. This was a year after I had left the program and a month before the actual launch. On October 7, 1998, I testified before the Space Subcommittee of the House Science Committee concerning the push to launch the *FGB* even though it was clearly going to have to wait much longer than planned before the *Service Module* arrived. Every indication was that the gap between the *FGB* and *SM* launchings wasn't going to be the intended four months, or even twice that. It looked as if it was going to be a long, long time before the *SM* got up there, and many in NASA had begun to doubt that it ever would.

Experience, I reminded the members of Congress, should have taught us that before committing hardware to space—a very hostile environment, full of unpleasant surprises—we should minimize surprises back on Earth. I urged that at the very least, the *Service Module* and other Russian support hardware must be certified as being "on track" by some independent evaluation. Also, the threat of *Mir*-related diversions must be ended, most reliably by the termination of that program. Such steps would take several months. Until such steps were taken, I considered it foolhardy to deliberately enhance programmatic risks—and our vulnerability to future blackmail—by launching the first elements, given such uncertainty.

I criticized what I called a "mindless momentum" toward an *FGB/Node-1* launch based on the same illusory hopes for future Russian support that year after year had been dashed. I urged that the *FGB's* launch date of November 20, only weeks away, be reconsidered, if not by NASA, then by those who influence NASA.

Using a football analogy, I argued that throwing the first two modules (the *FGB* and *Node-1*) into space now without knowing when, if ever, the planned follow-on hardware would show up to be linked to them was a hundred-million-mile long shot, the "longest 'Hail Mary' pass in history." One congressman laughed so hard he nearly fell off his chair.

But NASA's concern was to "get something in the air," both to inspire workers who were frustrated by years of delay and to show Congress and the public that there was indeed light at the end of the space station tunnel. Psychological and inspirational factors were paramount. So a module that had been conceived in symbolism would turn out to be delivered into space on a schedule that had symbolism as its primary motivating force.

The pass would eventually be caught successfully, but not without a series of heart-stopping bobbles. As everyone outside NASA had warned, the wait for the *Service Module* wasn't four months, or even six, as NASA had claimed that it might be in the "worst case." It was 21 months, longer than the certified design lifetime of the *FGB*. During that time, NASA had to perform three extra space shuttle visits to service the module's faltering power system, which hadn't been properly designed in the rush to launch.

On November 22, 1998, the *FGB*—the first piece of the *International Space Station*—was carried into orbit atop a *Proton* rocket. Another flaw in the design very nearly stopped the program dead in its tracks within hours of launch. Once the *FGB* had reached orbit, champagne corks popped at Baykonur, but at Russia's military space control center southwest of Moscow, it was eyeballs that were popping. As the *FGB* passed overhead on its first orbit, controllers radioed up some routine instructions for the autopilot to prepare to raise its orbit. The *FGB* sailed on, not responding. The commands were not even acknowledged.

The controllers at Krasnoznamensk (also called Golytsino) were accustomed to crisis. Military teams there controlled 120 of Russia's operational satellites, all but *Mir*. Most of these satellites had far exceeded their design lifetimes and were limping along, acting up, and even completely breaking down. But the *FGB* wasn't supposed to start off acting as if it were dying.

The *FGB* command system used a military radio link, called KOMPARUS. Russia never told NASA the command codes, and NASA eventually abandoned plans to develop independent monitoring, command, and control capabilities. During our orbital design activities, my team had again and again pounded on the Russians to at least disclose simple things that we needed to know. One of the most important was a list of the locations of the Russian ground tracking sites, so that we could compute the first and last opportunities for communication each day. It took a year to get a serious response, and when it did arrive, I doubted whether it was accurate. And in the year between my departure from NASA and the launch of the *FGB*, two of the sites either broke down or were shut down for lack of funds.

Understandably, there were no NASA observers at Krasnoznamensk. Probably, few NASA workers even knew that it existed, and none of them knew exactly where it was located. No Americans had ever been invited to inspect it. On the other side of Moscow, at the famous TsUP, a control room had been modified to monitor the *FGB*'s telemetry. But the raw data on the screens there came originally from Krasnoznamensk.

So on the day the *FGB* was born, and appeared to be about to die, a lieutenant colonel named Nazarov was facing the crisis alone. He was no stranger to stress. Along with the rest of Russia's military officers, he was severely underpaid (the equivalent of $70 a month), and his pay was years in arrears. But he was on duty.

The *FGB* was in its preliminary "parking orbit," just above the atmosphere. Without rocket burns within a day or two, it would start to tumble, then quickly fall back and burn up. A miracle was called for.

As soon as the *FGB* passed out of radio range, Nazarov began attempting a long-range diagnosis of the problem. The *FGB* passed over Russia twice more. Soon it would drift too far west to be contacted for almost a day. With one communications pass remaining, Nazarov formulated a theory.

Exactly how his intuition arrived at this approach was never made clear. It would be a year and a half before the Russians even disclosed that the event had occurred at all. But the idea probably came from a casebook of previous command failures of *FGB*-class vehicles. NASA was never shown this list, of course, and wouldn't have been able to do anything with the list even if it had seen it.

Carefully, Nazarov sketched out an alternative command code scheme that he thought the autopilot might accept. Ninety minutes later, the *FGB* had circled around Earth and once again appeared over Russia, for the last time that day. Nazarov sent up his modified command codes.

They worked. The autopilot saluted in electronic style, accepted the commands, and turned the spacecraft so that its maneuvering engines pointed in the correct direction. It moved farther out into space, away from the upper atmosphere. The spacecraft, and the entire program with it, was saved.

The Russians decided not to tell NASA about the scare. What NASA didn't know couldn't hurt it, they probably figured, not for the first (or the last) time. A software error in the *FGB*'s autopilot was soon confirmed (as Nazarov had guessed) and quietly corrected by radio commands from Earth.

It's for cases such as this that NASA develops backup procedures. With its worldwide tracking capability, NASA could have stepped in to tackle the problem if Nazarov and his team had failed. But the Russians made that impossible by not providing NASA with the command codes and never informing NASA of the crisis. In order to maintain control, they subjected the project to a near-lethal risk in the very first hours of its flight. And they got away with it: There's no evidence that NASA ever complained, even after the agency found out what a close call it had been.

For Nazarov, the incident also had a happy ending. He got a medal, a thank-you note, and a cash award of several months' pay, plus all of the back pay that was owed to him. It made him a very happy man, even though he had been the last man on Earth to celebrate the success of the *FGB* on launch day. In truth, he had been the first man to be justified in celebrating, since all of the other festivities took place only because the celebrants were not aware of what was really going on.

15

Follow the Money

"As scarce as truth is, the supply has always been in excess of the demand."

Josh Billings, Affurisms, 1865

"Time is precious, but truth is more precious than time."

Benjamin Disraeli

When I handed General Pyotr Klimuk the envelope containing $30,000 in cash, he barely looked at it before he dropped it on the table. He never even stood up. Such transactions seemed to be pretty routine for him.

"My clients ask that you sign this receipt," I continued. "You also agreed to sign this statement that you have counted the money and verified the correct amount." I had practiced my Russian to make sure that I got the legal terms correct.

Klimuk shrugged and took the page I offered him. Without a word, he glanced at it, placed it on the nearby table, and signed it.

Across the room, General Yuriy Glazkov was counting his own portraits of Benjamin Franklin. His envelope contained fewer than Klimuk's, a reflection of their pecking order. He signed his own receipt impassively. Neither thanked me.

These were two of Russia's veteran cosmonauts, with praiseworthy careers behind them. Klimuk had been the youngest Russian ever to command a space crew, and then in 1975 he had stepped in to lead a record-breaking space station mission after the original crew's rocket had crashed in the Altay Mountains. Glazkov had been on the space crew that had reopened another space station when it was feared that its atmosphere had been contaminated by chemicals or fire. The two had earned respect, and now it seemed only fair to them (and to others, including me) that they be allowed to cash in on their status.

Both men reminded me of fire hydrants: short and stocky, but hard-surfaced and tightly sealed. Together they ran one corner of the Russian space program, the Gagarin Cosmonaut Training Center at Star City. They had their fingers in all of the financial transactions going on within their reach. Whenever money changed hands—for training fees, for equipment purchases, for sales of memorabilia at auction, or for any-thing else, so it was said—they got their standard cut. Contacts of mine who worked there told me that they were universally feared and envied by those within their sphere of power. They apparently weren't very well liked either, even by each other.

Here I was in a motel room in Houston, engaged in a cash transac-tion with the commandant of Russia's cosmonaut training center and his deputy. They had just earned this money by allowing other cosmo-nauts to offer their own space souvenirs at auction in the United States. In return for giving this permission, they received a cut of the profits, and I was acting on behalf of the organizers of the auction to complete the deal. It was a routine "cost of doing business" in Russia, and one that was hardly limited to the space industry.

For decades, the nations of the world have engaged in space explo-ration for a wide variety of motives. For many, national prestige and military advantage were the main goals. Projects sometimes promoted diplomatic goals or advertised specific national technological capabilities. Scientific curiosity also counted from time to time, but usually not for much.

Only within the last decade or so has the idea of making money from space activities begun to appear. But while many of the old motivations still echo in the back of the minds of the space station program leaders, in an astonishingly short time, money has come to dominate all other themes. The Houston motel room was merely a precursor to its appearance. In

1998, for the first time, commercial corporations exceeded the world's governments in space expenditures. And many governments have actually become booking agents for the commercial sales of domestic space hardware and services. Even where the classical themes still weakly resound, they are often tolerated merely as a means to this new end, as camouflage or sweetener on top of some very mercenary transactions.

And without an appreciation of this new theme—without an attempt to "follow the money" into outer space—any efforts to understand and appreciate how we have gotten into such a dependent situation with Russia on the space station, and what we might be able to do about it, are hopeless.

Why had Russia repeatedly failed to meet most of its commitments to provide support for the *International Space Station*, and how can one explain the exceptional cases in which it has managed to meet them fully? Why did the aging *Mir* space station remain in use year after year, to almost three times its originally rated lifetime, at great risk to the Russian cosmonauts and any guests who happened to be aboard? And why has Russia performed triage on the varied components of its space industry, leaving some to wither, others to flounder, and still others to flourish?

Russia is currently earning almost a billion dollars a year from the sale of space services to Western customers ($880 million in 1998, and more the following year). This money, which accounts for 70 to 80 percent of the program's operating budget, has also allowed Russia's military space activities to avoid collapse. Western money has only leased (but not bought) the loyalty of Russian space industry leaders to the idea of international cooperation. Money dominates the politics and the technology of the Russian space program, especially with respect to our partnership in space projects.

Without an understanding of this fact, events appear to be a miscellaneous, aimless jumble of incidents and unpleasant surprises. But with an appreciation of this radically new (and still widely overlooked) theme in space activities, patterns become clear, current actions become understandable, and future courses become chartable.

Don't be distracted by the roaring rockets, the whirling centrifuges, and the zero-G soaring cosmonauts. Avoid getting fixated on the sprawling cosmodromes, soaring satellites, or blinking panels at Mission

Control. Don't be confused by the trackless expanses of empty space or the mind-numbing pages of budget statements. Follow the m-o-n-e-y, and all will become clear.

Of course, this journey won't be without hazards. And sometimes they get downright personal.

"Who told you about those houses?" the angry Russian asked me, poking me in the chest with his finger.

Taken aback, I could only say, "People I trust."

"Well, trust me," he replied, speaking very carefully. "It—is—dangerous—for you—to write—about—this subject. You know about the Russian mafia? It is dangerous for you to write about these houses." He walked off.

Vladimir Solovyov, a former cosmonaut and now a high official in his country's space program, had approached me inside the employee cafeteria at NASA's Johnson Space Center in Houston, Texas. It was early 1995, just before the first visit of an American astronaut to the *Mir* space station.

A few months earlier, the *Washington Times* had published one of my articles on the Russian-American space partnership. Along with my descriptions of the poor treatment of Americans in Russia and of NASA's poor knowledge of the Russian space industry, I had reported that vast amounts of Western money meant for the space program appeared to be winding up in the pockets of top officials, leaving ordinary workers severely underpaid.

In particular, I had mentioned the half-million-dollar mansions for top officials at Star City, the cosmonaut training center near Moscow. Reportedly hooked directly into the heating and electric power net of the cosmonaut center, these houses were being built only half a mile from the main entrance road to the base. And they were far too expensive for the salaries of the officials involved.

Now, unsure whether Solovyov's message was a threat or a friendly warning, I reported the encounter to NASA security. I never saw any results. For my own protection, I called Solovyov at his motel a few days later and asked him to clarify his warning, which he did. This time, for my own protection, I taped the conversation.

The joint space program continued, and in Moscow, construction of the so-called cottages of the "big cones" continued. "Big cone" (*bolshaya shishka*) is Russian slang for "bigwig" or "brass hat."

Two years after this encounter, by which time I had published another article on the subject that was illustrated with a photograph of some of the mansions, the tables were turned on the Russians. TV news reporter Byron Harris of the ABC affiliate WFAA in Dallas, Texas, had seen my articles and made some investigations of his own. He had been to Moscow and had found a way to visit and videotape the still-growing complex of space industry mansions.

When General Yuri Glazkov, one of the mansion-owning officials, visited Texas in early 1997, Harris was ready. Harris, his interpreter, and his camera crew entered the Johnson Space Center and lingered outside of administrative headquarters, where Glazkov was headed for a meeting. As the Russian and his interpreter approached, Harris asked him about the funding for the costly dwellings.

"There are no such houses at Star City," Glazkov answered through his interpreter. Asked about his mansion, he kept up his bluff: "I don't know anything about it, I don't live there." Confronted with the evidence that he did, and that one of the houses was in fact his, he changed his tack: "This has nothing to do with the space program, and I don't want to talk about it." Harris asked where he had gotten the money. "My wife is a pilot, and we have saved up all our lives for this," Glazkov explained, walking off. "I can have such a house if I want to," he added.

Of course, any money that Glazkov and his wife could have saved would have been wiped out by the hyperinflation during the collapse of the USSR. His official salary was totally disproportionate with the value of the house in question.

NASA's reaction told whose side the agency was on in the conflict between potential Russian space corruption and the American public's "right to know." The agency immediately clamped down on the U.S. news media. Highly restrictive new badging procedures for journalists were immediately implemented, to make sure that no visiting Russian space official ever had to go through such an embarrassing interrogation again.

When Harris's report appeared on *Nightline* a few weeks later, a NASA spokeswoman gave the agency's official position on the mansions: "What Russia does with their own money is their own business." Pressed as to how she knew that this was really Russia's own money and not assets diverted from Western-funded programs (as is commonly believed by workers at Star City, foreign and Russian alike), the official admitted that she had no idea.

These profits were enormous, because Western governments had actually required the Russians to overcharge for their rockets so as not to compete too vigorously with foreign rockets. The Russians obliged, and doubled their asking prices. The torrent of windfall profits poured into infrastructure improvements, private bank accounts, or just disappeared.

NASA made the not unreasonable point that the Russians (particularly those at Star City) were in fact delivering on all of their contracted services. If the money wound up as bonuses for top officials, it only gave them salaries commensurate with those of Western space big-wigs. This was just the way things were handled in modern Russia.

But I made sure that Congress heard that there were plenty of other things about our Russian partners that NASA simply didn't want to see. In a prepared statement that I wrote for the congressional hearings on October 7, 1998, I gave examples. NASA had made certain that American decision makers would not be distracted by evidence of corruption within the Russian space industry. Regarding the notorious cosmonaut mansions at Star City—which some White House experts still blindly dismissed as merely "allegations"—within NASA, it was a strict rule *not* to mention them. I recounted how when one NASA official was outraged enough to describe the mansions in a trip report, he was ordered to delete the mention of the mansions and resubmit the report. Other NASA workers at Star City told me that it was made clear to them that any overt interest in the houses would be severely career-limiting. Such a policy made it easier for higher officials to act surprised and incredulous when confronted with independent evidence for such a diversion of funds.

One member of the subcommittee, Eddie Bernice Johnson (D-Tex), was courteous enough to send me a detailed written response. After addressing several other points with thoughtful responses, she brushed aside my mention of the houses: "Regarding your comment of 'the mansions being built at Star City,'" she wrote, "I have been assured by the Administration that there is no factual basis for this assertion." That was that—the White House said that the houses weren't there.

But eventually even the Russians began to acknowledge the out-of-control corruption within their space industry. In April 1997, at a briefing by the economic crimes unit of Moscow's Internal Affairs Main Administration, specialist Timur Valiulin described a wide-scale pattern of top-level corruption in the aerospace industry. He singled out the Lavochkin Bureau, where the *Mars-96* probe had been built (it crashed in November 1996 after tens of millions of dollars of European investment). He described how the bureau's general director and one senior associate had been arrested for embezzlement prior to the probe's crash. He called the case "merely an individual fact in a series of such outrages." Shortly afterwards, the Russian government fired Oleg Soskovets, the point man for negotiating the U.S.-Russian space partnership since 1993. There were accusations of massive personal corruption.

During an unrelated investigation of banking irregularities in Moscow in 1999, officials opened and examined the contents of the safe deposit boxes at one particular bank. In one box they found a paper sack containing $1 million in cash. The box's owner, a top official at the Khrunichev Institute, was asked about the money. He provided an acceptable explanation, and it was returned to him.

The money for space flight was only a fraction of the Western cash that poured into post-Soviet Russia. It was supposed to ease the country's transition to democracy and a free market, but in hindsight, most of it appears to have gone to the oligarchs, a group of those who had run the system before and now owned much of the new industry. It was also supposed to keep otherwise unemployed Russian rocket experts from assisting weapons development programs in rogue states around the world.

But RSC-Energia is a good example of what really happened. This firm builds and operates Russia's human piloted space vehicles and thus plays a crucial role in the *International Space Station*. The corporation was privatized in 1994, with the government owning 51 percent of the stock (now down to 38 percent). Its 1997 commercial earnings were placed at $350 million, about half of the total foreign sales of the entire Russian space industry. That broke down to:

- $160 million for foreign guests aboard *Mir*, mainly American but some French
- $100 million for the sale of space station hardware, some paid for by NASA and some still owed by the Russian Space Agency

- $50 million from investment in Sea Launch, a joint project with Boeing to launch a Ukrainian-Russian rocket from a ship in the Pacific Ocean
- $20 million from the sale of a commercial third stage used on *Proton* rockets for Western satellites
- $20 million from the sale of the Yamal communications satellite to a Russian bank

Yet during the budget cutbacks of the late 1980s and the collapse of the early 1990s, Energia was forced to lay off more than 40,000 space workers. The leaner space firm now has 22,000 engineers and technicians. It scrupulously adheres to export controls regarding technology transfers to third parties, but it has no control over the workers it was forced to lay off.

Between 1991 and 1994, according to the Moscow newspaper *Trud*, 115,000 engineering and technical personnel and 90,000 industrial workers left the aerospace industry. Yuriy Koptev reported that if current trends continued, only 100,000 of the original 360,000 people in the space industry would still be on the payroll by the end of 1995.

At the *Progress* plant in Kuybyshev, where *Soyuz* boosters were built, by early 1995 "only true patriots have remained," according to Vladimir Pishistov, production unit chief. Plant director Anatoliy Chizhov told a TV news program that wages were not high enough and were not being paid on time, and he urged a return to the "old values." By early December 1994, wages had been paid only through September, and 8000 workers had already been laid off. That same month, rocket engine builders at the Glushko plant Energomash announced plans to strike, since they hadn't been paid in two months.

In a more recent interview on the space.com Web site, Valery Timofeyev, the first deputy general designer of the NPO Lavochkin Center, which builds interplanetary probes, discussed the demographic crisis that he was facing. It was typical of the overall Russian space situation. Employment was down to 5500 from 11,000 workers 10 years earlier. The average age of the workers was about 50. "Young people are not interested in working for the NPO," Timofeyev lamented.

Forty years ago, when *Sputnik* and *Vostok* went up, Moscow selected the best and the brightest young graduates from its engineer-

ing and scientific institutes to staff its grand new enterprise, space exploration. For the most part, those same men and women remained at
these jobs for their entire careers. And since there weren't any additional staff increases, there were very few "new hires," the young engineers who are the future of any technical enterprise.

Now, at the turn of the century, Russian space veterans (most of
them in their sixties) are retiring or dying off at a very high rate. "Because
persons specializing in space activity are not receiving the material support they need, an irreplaceable exodus of highly experienced personnel
is occurring," stated a letter from space veterans to Boris Yeltsin in 1997.

Without funding to attract a new generation who can pick up corporate knowledge from the old-timers, Russia's space engineering experience will die out. And if this happens, any projects funded in future,
more prosperous times will have to literally start over again in training
personnel.

But in the meantime, Russian space workers faced more immediate
concerns. One senior manager at Energia told me in 1996 how he had
to pay the 200 workers in his section. "They give me a pile of cash for the
payroll," he began. "I count it and sign for it, and it's never enough." The
money is intended for back pay, often months in arrears, and the pay
rate—perhaps $200 per month—is half of what taxi drivers earn. "I then
must call in each of my employees one at a time," the official continues,
"and we negotiate over how much money they really need to get by in
the next two weeks." Sometimes he had money left over at the end;
sometimes he ran out before everyone was paid.

Energia is still in business and still produces first-rate space hardware. But many less fortunate firms, having lost their government military contracts and being unable to find Western customers, have
collapsed entirely.

During a NASA tour of Russia's Mission Control Center in 1994,
the Americans saw no sign that the NASA cash transfers were being
spent on facilities or payrolls. "The [Energia] guys said that RSA skims
almost all of it off before they see any," one of the flight directors later
told me.

"They are so desperate for money, they try to charge NASA extra
for everything," he continued. "There are studies for STS bringing water
supplies to *Mir*, or for transferring fuel cell waste water onto the *Mir* for

drinking, since water supplies are so tight [it's normally dumped over-board from the shuttle]. After a technical meeting one of the Russian officials asked his NASA counterpart which NASA contract would be used to fund the Russian studies of modifications to their water supply system. These were modifications required to allow the U.S. to give them—gratis—critical water supplies! Unbelievable!"

In 1994, NASA officials decided to provide the American space vis-itors on *Mir* with American space food. But the adverse psychological stresses of forcing the Russians to live on their own space supplies, side by side with the American visitor's sumptuous cuisine, prompted NASA negotiators to offer to provide food for all three crew members. The Russians accepted the offer. But since the additional weight of the free food exceeded the previously agreed-upon NASA baggage allot-ment, the Russians submitted a bill to the American side for excess transportation fees!

Such financial challenges also created other personnel problems. In December 1997, an ABC-TV news reporter interviewed Viktor Blagov, head of the Mission Control Center in Moscow. "Are you finding it diffi-cult to keep highly trained technicians?" he asked. "Do you have brain drain in your space program?"

"Yes, we have it," Blagov admitted. "We have another unpleasant trend—the ageing of staff. It is affected by the inadequate salary, and seri-ous temptations to earn good money elsewhere." But Blagov shared some good news, which was made possible by the increasing Western aid. "On December 1, two days ago, our salaries were doubled, that will help us greatly to keep our staff."

White House spokespersons and American aerospace industry leaders have repeatedly raised the bogeyman of renegade unemployed Russian engineers going to work for rogue states such as Iran or North Korea to help them with their military missile programs. According to this thesis, the billion dollars a year of Western money that was going to Russia for space applications would prevent this.

While this was a laudable goal, it soon degenerated into a vain hope. Today it can only be considered a dangerous delusion. Here's why.

First of all, hundreds of thousands of Russian rocket engineers—those who were most closely connected with building surface-to-sur-face missiles—had already been laid off in the early 1990s as the Soviet

Union's massive ICBM program collapsed. Only about half remain at work in the much-constricted factories.

Second, the overseas market for "renegade rocketeers" has always been tiny—perhaps no more than 100 or 200, usually a lot less. That was all the rogue states could afford, or even needed for their own programs. And they tended to be short-term contractors, brought in for a few months to address specific engineering bottlenecks, then sent home when the problems were overcome.

Russian journalist Yevgenia Albats interviewed one such engineer at an airport in Moscow in 1998. He was off to Iran for two months, and he reported having no difficulty in making the trip. He was being paid only $260 per month, or little more than the salary of a rocket industry worker inside Russia. As with any officially forbidden commerce (such as illegal drugs in the United States), when the price is relatively low, as it is here, this shows that interdiction has been a failure.

The official U.S. pretense that subsidizing half of Russia's rocket workers, while leaving the other half unemployed or underemployed at menial tasks, actually constrains such arrangements is just silly. It's like trying to control the birth rate of some wild pest species such as rabbits or coyotes by eliminating only half the males. Any naturalist can tell you that the remaining ones are more than enough to take up the slack. Whether it's a case of breeding in the wild or hiring renegade rocket scientists, poorly designed restriction schemes have always been totally ineffective.

Congress asked Deputy Secretary of State Strobe Talbott, one of the chief architects of the Russia policy for the Clinton administration, about these issues. Specifically, James Sensenbrenner of the House Science Committee asked whether the United States' inability to make the Russians perform at promised levels on the *ISS* might affect our ability to make them stand by other agreements, such as nonproliferation. In a written response dated October 7, 1998, Talbott replied: "We believe that the participation of the Russian space industry in the *International Space Station* gives Russian firms a strong incentive to forego cooperation with problem states in sensitive technology, and thereby supports our nonproliferation efforts."

Talbott insisted that things were getting better: "Over the past year, Russia has taken important steps in this area. The Russian government has implemented a catch-all authority and other mechanisms to help it ban all cooperation with missile programs of concern. It also has stopped some transactions, but has not ended cooperation with missile programs in countries of proliferation concern. We plan to continue to work with the highest levels of the Russian government to resolve this problem."

A year and a half later, in March 2000, the U.S. ambassador to Russia, James F. Collins, admitted that "there are definitely still problems" in the area of technology proliferation, although he stressed that the companies doing business with Western partners were not involved. But that's only half of the available population of rocket experts.

Collins's comments were in response to a February 2, 2000, CIA report to Congress. It stated that "despite some examples of restraint, Russian businesses continue to be major suppliers of weapons of mass destruction equipment, material, and technology to Iran." And, as Albats discovered, a vast pool of Russian experts remain "for hire" as individuals.

Throughout the 1990s, it was a buyer's market for Russian rocket expertise, wrote David Hoffman, the *Washington Post*'s Moscow correspondent. Viktor Vyshinsky, head of a Russian space-flight research facility, told Hoffman that there was little to stop a scientist from leaving Russia. "The only thing that stops you is scruples," he said. "But if someone takes it into their head to sell something, then I don't think there will be a problem."

Official controls by the Russian government were hard to detect. "I do not know of any major cases of prosecution of export control violations which put people in jail," said Vladimir Orlov, director of a Moscow nonproliferation group. Hoffman quoted official estimates that about 60 percent of the core scientists have been helped by Western funding. "But where the rest have gone is simply not known," he wrote. "Nor is it known how much technology and know-how slipped out of Russia."

The bottom line is that Western money has been able to buy just about anything in the Russian space industry, and in most cases it has been the *only* way to acquire goods and services there. There is a frightening corollary to this, which is that if Western money can buy anything, so can rogue state money. The idea that Western funds have bought us exclusive access to Russia's space and rocket know-how ranks as one of the most dangerous delusions of the 1990s.

16
ISS Opens

*"For a successful technology, reality must take
precedence over public relations, for Nature cannot be
fooled."*

Richard Feynman

*"The chess board is the world, the pieces are the
phenomena of the universe, the rules of the game are
what we call the laws of Nature. The player on the
other side is hidden from us. We know that his play is
always fair, just and patient. But also we know, to our
cost, that he never overlooks a mistake, or makes the
smallest allowance for ignorance."*

Thomas Huxley

As the empty two-module *International Space Station* approached
its first birthday in November 1999, space workers prepared a
special party to celebrate. They planned to launch the next piece of the
station, the *Zvezda Service Module*, on November 12, 1999.

Yet the rush toward the launch date was probably the greatest
example of a space bluff in the history of an international space project
known for bluffs and balks. Both sides knew that the date was unrealistic, but neither would be the first to admit it. It was like two stubborn
head-on drivers in a game of "space chicken."

The NASA inspectors who visited the payload processing facility at Russia's Baykonur Cosmodrome reported that the *Service Module* was indeed nearly complete. True, some critical equipment, such as a rendezvous guidance radar, wasn't scheduled for installation until January, and true, flight software for the German-built control computers had not yet been delivered, but still, the November 12 launch date was "realistic."

The *"Service Module* is ready for November" charade went pretty far. In October, the press office at NASA was even requiring the news media to file their formal requests for credentials to cover the phantom blastoff. This is a step that is usually conducted about 30 days before a flight. The two sides were eyeball to eyeball, and nobody was blinking.

In the end, bad luck provided a way to save face: Another Russian rocket blew up, and all further launchings were grounded.

Russia's most powerful rocket, the *Proton*, had been slated to carry the 20-ton *Service Module* into orbit. With an upper stage, it could also carry communications satellites into 24-hour synchronous orbits. For several years, it had been doing this very profitably for Western customers.

Then, on July 5, 1999, a *Proton* rocket's second stage caught fire and exploded during launch. After several months of investigation, engineers identified a contamination problem in the engine turbo pumps, fixed it, and resumed launches. Two successes followed; then yet another *Proton* blew up on October 27, in a repeat of the July 5 accident.

The *Service Module* launch, little more than two weeks away on official calendars, suddenly had a respectable rationale for cancellation.

A more thorough accident investigation revealed that the engine turbo pump problem was more severe than had been realized at first. A faulty batch of engines was identified and isolated, and commercial launchings resumed in February. But an entirely new set of turbo pumps had to be manufactured for the Russian program flights, and this put off the *Service Module* launch until July 2000.

And the launch worked. To the relief of space workers in both Russia and the United States, the *Service Module* and its life support hardware were successfully linked to the *International Space Station* on July 26, 2000. Within two weeks, a robot supply ship docked to the still-unoccupied station. This was followed on September 12 by a week-long unpacking and outfitting expedition by seven space shuttle astronauts and cosmonauts, who were sent up to ready the station for its permanent crew.

At long last, the Russians had performed—and, as hoped, the *Service Module* was a beautiful piece of work. But now it was NASA's turn on center stage. Using the space shuttle, it had to perform a series of three critical delivery missions, designated 3*A*, 4*A*, and 5*A*, to the now-livable station. In parallel with the first delivery mission, the *ISS*'s first permanent crew was sent to the station aboard a Russian *Soyuz* spacecraft on October 30.

As with any highly complex project, different kinds of problems continued to pop up late in the game. But battle-scarred NASA managers took them all in stride. "In a program this big, hardly a week goes by that we don't have a situation we need to work on and make sure that everything is going to be OK," explained Tommy Holloway, NASA's space station program manager, at a press conference in Houston on September 26. "At this time, we don't have anything we would classify as a show-stopper."

Others were less upbeat. For example, even optimists admitted that the development of the computer software for the space station had been difficult. Software glitches in the hundreds cropped up, so many that pessimists believed that NASA should have considered delaying the missions. If NASA pressed on, a friend warned me, "we may very well spend a lot of time putting our fingers in an ever-increasing number of holes in the dike." Another told me that in his experience, complex software packages usually work well at first but then suffer a serious crash after about six months of operation.

One problem led to another. Inadequate software led to the late delivery of training simulators for both the ground teams and the astronauts. Without the software, NASA's plan to test the space station hardware as thoroughly as possible on the ground could not be fully carried out in time for the planned launches. Such testing had already revealed many potential problems in the equipment in enough time to fix them, circumvent them, or learn to live with them. But since each test uncovered still more problems to be fixed, experienced engineers knew that there were a whole lot more to be discovered.

Of course, the space station itself was designed to have backup systems, margins, and reserves for unexpected stresses. Above all, NASA wanted the crew to have the flexibility to work around last-minute problems manually. That last feature is one of the frequently advertised key benefits of a space vehicle with people aboard.

NASA planners were counting on trained crews and work-around procedures to overcome known deficiencies in space station hardware and software. However, the crews were less prepared for the expected challenges and the procedures less worked out than at any other time in the history of U.S. human space flight. This was because the high-fidelity simulators were late.

"We launched the space shuttle when we were 90 percent ready," one veteran space worker told me, "but we're launching the space station at only 50 percent."

The consensus of the working troops, however, was that further waiting would have been unlikely to reduce the risks. It was time to do a shakedown under real flight conditions. There was one saving grace to the space station that made it sensible to fly under such conditions: Because of the orbital stability of the space station and the presence of the fully functioning Russian modules, the actual threat of vehicle or crew loss as a result of hardware and software problems was lower than on any previous U.S. astronaut mission. Frequent problems, including major failures, were expected during the many months of assembling the station, but experts believed that this was an ordeal that had to be faced. Waiting would not make it any easier.

NASA's three big missions were to add to the three modules already in orbit. The Russian-built, U.S.-financed *FGB*, code-named *Zarya*, had been launched in November 1998. Its solar arrays provided electrical power for itself and the U.S. *Node*, and its rocket engines provided thrust to keep the complex pointed in the right direction and at the right altitude. *Zarya* had been repaired, serviced, and stocked with supplies by three subsequent visiting space shuttle missions. The U.S. *Node-1* module, code-named *Unity*, had been carried up on a space shuttle mission in December 1998. It had six attachment ports (one at each end, four around the waist) for mating with other sections. And the Russian *Service Module (SM)*, code-named *Zvezda* and launched in July 2000, would provide life support for the crew members who were to remain permanently aboard the complex beginning in November.

For NASA, the year 2000 was the Year of the Big Push—several big pushes, in fact. These were labeled 3A, 4A, and 5A, and they were brought up aboard successive shuttle missions. In each case, shuttle astronauts conducted space walks while the shuttle's robot arm moved the equipment around and hooked up the new sections. *Mission 3A* carried what was called a truss assembly, 4A carried the solar power wings, and 5A carried the pressurized *Laboratory Module*. If all these components could be made to work, the station would reach its intended capability.

As the first crew reached the station and the big push approached, I did a technical assessment of the challenges for *Spectrum*, the magazine of the Institute of Electrical and Electronic Engineers. I compiled a list of the most serious problems at that stage of the project. Interestingly, they tended to be electrical or electronic in nature:

- Electric power shortages caused by the last-minute need to run heaters to warm electronic components that were vulnerable to cold
- Concerns over the danger of electric shock to space-walking astronauts because of the failure of the device that equalizes voltages between the station and the surrounding space plasma
- Anxiety over the unexpected degradation of crucial optical-fiber data lines
- Frustration with very late and error-filled control software
- Concern over inadequately trained personnel, both on the ground at Mission Control and in space, and over poorly checked flight procedures

These specific issues provide fascinating insights into the degree to which the NASA team was chasing last-minute glitches and patching last-minute holes. In the end, it also shows how successful the team was in finding and fixing such problems.

For each hardware and software problem, NASA and its prime space station contractor, Boeing, had developed backup techniques. They expected that these techniques would allow the station and its crew to function safely and effectively or to recover smoothly. But they found that repairs are often made at the cost of breaking something else.

For instance, a repair made on the control moment gyroscopes (CMGs) had some unintended consequences. Key components, CMGs are needed to orient the space station—that is, to control its attitude in space. By using electric motors to vary the speed and orientation of massive flywheels, control computers can cause the station to turn in the desired direction without the use of steering rockets. Four of these CMGs were included in the *Mission 3A* package, launched in mid-October, that brought up two midsized space station sections. But they weren't activated until the full computer suite arrived with *Mission 5A* the following February.

During low-temperature testing of similar hardware earlier in 2000, the sensor that measures the flywheel spin rate had failed. The engineers had concluded that the epoxy mounting of the sensor had cracked. Replacing these components of the flight hardware might have taken more than a year, because an improved version would have to be developed and tested. And such a long delay was unacceptable.

Instead, NASA opted to raise the operating range of the thermostats controlling the electric heaters in the CMG units. This would keep the hardware much warmer than the level that had been fatal in the initial low-temperature tests. And later testing indicated that the malfunctioning unit had been much more susceptible to thermal stress damage than the higher-quality flight-qualified units that were manufactured subsequently.

The electric power budget for the station in the "post-3A" interval was already extremely tight. The new U.S. hardware would have to depend on the solar arrays of the Russian segments for all of its power. It turned out that if NASA allocated enough power to run the heaters to keep the CMGs safe, there would not be enough power to heat the *Node-1* module. Unheated, that module would be exposed to water condensation from the breath and sweat of the astronauts, a situation that could damage its electrical components.

Consequently, when the permanent astronaut crew arrived on November 2, they found the U.S. *Node-1* module sealed off. They were restricted to the more cramped, though still adequate, living quarters in the two Russian modules, the *FGB* and the *SM*.

Minimizing the impact of closing the U.S. *Node-1*, NASA experts admitted that the power situation on board the *ISS* had already been too tight. "Even before the CMG problem, we probably didn't have the power to keep the *Node* open," astronaut Robert Cabana, now director of space station operations, told me for an article I was writing for *Spectrum* magazine.

Still, adding the CMG heater requirements made the tight power budget even worse ("severe power concerns" is how one NASA expert described it). If there had been further failures of the recently replaced nickel-cadmium batteries in the *FGB* and the *Service Module*, there could have been brownouts throughout the whole station. Flight experience had shown that a high failure rate could be expected, and, in fact, a second *Service Module* battery failed shortly before *Mission 3A* was launched. It might even have become necessary for the crew to power down the *FGB* entirely and retreat into the *Service Module* until *Mission 4A* brought up the U.S. photovoltaic arrays that would fix the power deficit.

"We will need to have good FGB batteries operating and spares when the crew arrives," a NASA planning memo stressed. But extra batteries had been needed before.

"One of the biggest challenges with the FGB is keeping the batteries functional," a top space station operations official told me earlier that year. "NASA thinks the failure rate is high, but during [the shuttle-*Mir* missions in 1996–1998] we delivered twenty-six or twenty-seven 800-A batteries and fifteen [power controller] modules to *Mir*, so the battery failures shouldn't be as much a surprise as they are. It's just that the people who are working the related issues within the space station are not the same people who worked the issues on [shuttle-*Mir*]."

Meanwhile, the ripple effect continued, and CMG-induced solutions created new problems with their own costs. A NASA Mission Control specialist reminded me that the notorious noise problem in the Russian segment (noise levels that could reach physically damaging ranges) was tolerated because "the crew would always be able to retreat to the *Node* and get away from it." Thus, the solution to one problem invalidated the previous solution to another problem. Now, at least for the time being, the *Node* would be unavailable as a quiet refuge.

Space vehicle designers generally have preferred to use power systems that operate in the 24- to 28-volt (V) range. But because the space station would use so much more power and would be so much larger physically, NASA engineers decided to operate the buses from the solar arrays at a heftier 130 to 180 V. This option reduced the current required, the wiring, the component weight, and, ultimately, the cost.

But the designers soon realized that they had created a new problem for the station. Orbiting Earth, the station flies at 18,000 mph through a very thin plasma, not a total vacuum. Since the power system was grounded to the station's structure, the outer skin of the station developed a relative potential of 100 to 160 V with the surrounding plasma. According to a NASA training manual on the station's power system, "40 to 60 Vdc has been observed to be the minimum required for arcing [sending sparks through the air]." With its outer skin at 100 to 160 V, the station would be way over that threshold.

Such a high voltage would have resulted in continuous "mini-arcing" across the skin of the station. This would damage the station's thermal coating and its viewing ports, and it would also degrade the solar cells. More dangerously, it would create a shock hazard to crew members on space walks.

To resolve this unacceptable situation, NASA built a suitcase-sized Plasma Contact Unit (PCU) to ground the station into the plasma by shooting a stream of ions into space. A hollow-cathode emitter fires ionized xenon; each emitter operates at 435°C and is surrounded by a protective metal cage to prevent contact with space-walking astronauts. In case one of the units malfunctions, there is a backup available: Two 170-kg units were carried on *Mission 3A*. But they wouldn't be needed until the high-voltage solar arrays from *Mission 4A* became operational.

Alternative methods of controlling the buildup of charge were investigated, a NASA reference manual stated, "but would require major Electrical Power System redesign." Building the PCUs was "the most cost effective method"—or so it seemed at first.

A few months before launch, a NASA safety team reviewing this solution discovered some faulty thinking. What would happen, the team asked, if the operating PCU were to break while the astronauts were outside the space station doing assembly or repair work? Analysis revealed a shocking answer: Without a functioning PCU, the charge differential could exceed the arcing threshold within seconds. A space-suited astronaut would act like a lightning rod, and he or she could experience a potentially death-dealing full ampere of induced current through the suit and into his or her body.

To repair the faulty "fix," another tiger team of specialists looked for procedural workarounds that would prevent such a hazardous situation from developing. The plan that they came up with would require that both of the PCUs be turned on before every space walk; NASA engineers verified that this was technically feasible. Then, if one PCU broke, an emergency procedure would shunt the high-voltage solar arrays off-line and turn them edge on into the space plasma flow, reducing the voltage differential below the arcing threshold. The crew would then reenter the station as quickly as possible.

But the engineers then discovered yet another problem that would keep this workaround from working. A breakdown of the PCU was not automatically annunciated to the station's computer or to Earth. Some software changes would be needed to add such a capability.

And there was another catch. The standard procedure for dealing with the failure of a PCU was a space walk to replace it with a spare. This workaround plan required both PCUs to be functional before the space walk begins, a catch-22 in orbit.

But the planners then developed yet another set of procedures. For a space walk to replace a failed PCU, the solar arrays would be taken off-line before the crew exited the station. The station would rely on battery power and the 28-V solar arrays on the Russian modules until the astronauts got the second PCU installed. They would then have to reenter the station while the new unit went through a days-long conditioning process to activate its hollow-cathode emitter.

These plans were accepted as the means of living with the deficiencies in the existing voltage differential mitigation hardware. The moral of the story is not that the equipment was faulty, it is that any highly complex system is always going to face such unexpected problems. The key to success is finding them soon enough to apply last-minute fixes with flexibility and imagination, which is what the *ISS* team was able to do in the final months before flight. They even added a small charge sensor device to be placed at the top of the station's electric power tower, as well as an extra space walk for the crew to hook it up.

Early in 2000, word began spreading of serious problems with the optical-fiber cables within the U.S. *Lab*, the keystone of the space station, which was to be delivered on *Mission 5A*. The fiber system is one of the three separate communications systems that transmit payload data within the module. The high-rate (100-Mb/s) fiber lines route the science data to downlink antennas or to other payload instruments. A low-speed line allows the crew to perform control and monitoring functions, and a medium-rate Ethernet line lets the crew monitor selected subsets of real-time data.

Trouble started when the optical-fiber lines began degrading, becoming more and more brittle as a result of the damaging effects of insulation outgassing, one expert told me. Another payload engineer told me that the flaws were caused by etching prior to the application of the carbon coating that was necessary to bond to the inner sheath. (It was found that just coiling the fiber during the sheathing process would also break it.)

To help solve the problem, Boeing developed a way to replace a damaged fiber bundle with an undamaged bundle in orbit. Doing this while *Lab* was still on the ground would have taken too long, and ripping out the existing fiber, replacing it, and following NASA test procedures would have cost too much. Another expert reported that the time required to replace and recheck all the fibers was estimated at 18 months. Instead, only those that actually broke during the stressful rocket ride into orbit would be replaced. Replacing some of the fiber bundles on the *Lab*, however, involved depressurizing the *Lab*, another potential hurdle.

Initial rumors regarding the fiber optics were overblown, and the problem has been resolved, NASA officials insisted. "We replaced some cables that we found had broken," NASA's Tommy Holloway told me before the launch. "They had been abused, bent too much, beat up during installation."

All of the lines had been thoroughly tested, Holloway reported, and they were fully functional. He confirmed that there had been "concern" over the unexpected degradation in the lines during fabrication. "We've recreated the process of building them, and during sheath application some chemical exposure reduced the tensile strength of the lines," he said. "There was no impact on transmissivity. We're now confident it's not going to deteriorate any more, so we're ready to launch."

The computer architecture for the U.S. side of the space station involves a multitiered array of distributed computing functions, with crew interfaces provided by plug-in laptops. One set of laptops (IBM 760 Thinkpads) supplies the crew with displays of the station status and relays their commands to the main computer network. A second set gives them access to the library of crew procedures and schedules, the inventory management system, and word processing.

When *Mission 5A* carried the U.S. *Lab* up to the space station, it brought more than a dozen computers (based on an Intel 386 SX chip) and about a million lines of code. The computers do practically everything except for locking the hatches and flushing the toilets. They automatically run the power generation and distribution system, the station's attitude control in space, the station's environmental control systems, and communications management. At the same time, they monitor the status and safety of thousands of components.

Consequently, the threat of software errors can be significant. It is not merely a matter of the system not doing what it's commanded to do or not exchanging the required information. Sometimes the system can do things that it was never ordered to do.

In an error discovered less than a year before flight, the corruption of two adjacent flags (bits in a status word) commanded an air valve to open while locking out the "valve close" command. Only a power cycle could reset the system and prevent all the air from leaking out of the station. In another dangerous flaw, a pressure-monitoring routine failed to recognize a normal airlock depressurization for a space walk. Assuming that the whole station was leaking, it issued orders to dump emergency oxygen into the modules.

The laptops themselves were not immune to problems. Although hitting any key was supposed to wake up the laptops from the quiescent mode, hitting the space bar actually froze the display and required rebooting. "There are so many software discrepancies, it's a challenge to list them all and make sure they're in the procedures documents," an astronaut expert on station software told me. Flight crews, he added, were concerned with the large number of software workarounds and limited testing in high-fidelity environments.

The distributed nature of the system required integrated testing, as a 20-year NASA veteran pointed out. "Once the integrated test is complete, a change to code in any single box invalidates all the code interfaces," he said. Yet the best that could be done was to test only the parts of the code that had been directly modified, even though there were at least 700 major changes in the U.S. *Lab* software.

NASA project officials remained optimistic. "The software is quite mature and is being exercised with the hardware on a near-continuous basis," former astronaut and current Boeing executive Brewster Shaw told Boeing employees early in 2000. "The problem report rate is significantly reduced." (Before and after figures were not disclosed.) And another expert told me that "while the software is not perfect by any means and the sheer number of workarounds and software deficiencies is mind-numbing, it should not delay the launch of the U.S. *Lab.*"

Other experts assured me that the software was finally good enough to launch. "In a program this size, it's just a fact of life it's going to go up imperfect," one said. Software problems will be fixed through patchwork upgrades over the period of at least a year. The first crews have trained with long lists of existing flaws and with advisories about what processes will probably not be working correctly.

The *Lab* module was hooked up in February 2001, and the computers on board were activated by the station crew. The computers began exchanging data with the Russian segment's computers and began sending control commands to the CMGs (from *Mission 3A*) and the solar arrays (from *Mission 4A*). Special software "tiger teams" watched the whole process, and they held their collective breath at each step.

While there were certainly enough "ankle-biter" glitches to keep them busy, at first the software troops encountered no significant problems. All of the major functions worked more or less as planned. And the last-minute fixes (some of them even entered manually after the mission had been launched) did their jobs.

One software list of equipment to be automatically turned off during a power reduction had to be modified so as not to unintentionally turn off the PCUs. While the U.S. team was using its new computers to operate a berthing mechanism, the Russians unexpectedly turned off their computers to load some new software of their own. This caught everyone by surprise, but the U.S. *Lab*'s computerized fault detection system noticed the shutdown and properly changed its own operations to stand in for the Russian signals. There were also some small problems with laptop interfaces, with timing synchronization signals, and with certain status flags that were swapped between computers. But "very happy" was the term used by the software teams to describe their feelings about what they correctly saw as "a monumental milestone for the program." These were the worst problems to occur during that month.

The first major software crisis didn't occur until April, two months later, when new software for the Canadian robot arm was added to the mix. At that point, a cascading collapse drove the station's network of computers to its knees, but the station survived and gradually recovered.

The previous year, difficulties with the software had created other problems. The delivery of error-filled control code had been late, affecting much more than just the space station. In a NASA status report in August 2000, the preparation of the mission operations team for *Mission 5A* was listed as *red*, in the words of the report, a cause for mandatory repairs. The low rating was given because of a "cumulative risk of flight controller certification threat [each flight controller must pass a strictly defined training plan involving complex simulations and study of software documentation]." Added to that was "the lack of an adequate multi-segment simulator for flight controller training and verification of procedures requiring a high level of integration."

For more than 20 years, NASA flight controllers have trained effectively for shuttle and *Spacelab* missions using sophisticated computer models that feed simulated space-flight data to the real Mission Control Center consoles in Houston. The simulations are directed by training personnel who script various contingencies and failures in order to test the flight team's ability to detect and recognize the problems and react to them properly.

Flight controllers (I was one for many years) subscribe to the old Chinese maxim that "the more you sweat in peace, the less you bleed in war." The Russians have a similar proverb, attributed to Kutuzov, the Russian general who beat Napoleon: "Training is hard so fighting is easy." But in describing the last-minute attempts to hook a space station simulation into the existing setup, more than one NASA veteran was extremely uncomplimentary.

"The [space station] software is extremely 'brittle,'" one expert explained, and it was hard to tell whether it was real or just a problem with the simulator. In space station stand-alone simulations, he said, the training team "can often put in just one well-placed malfunction at the start, and watch the whole station fall apart while the flight control team flails [NASA jargon for struggling ineffectively]." In fact, some teams never figure out how to solve the simulated problem. Because of that track record, during joint shuttle-station simulations, "the [training] team is often scared to put in any malfunction at all" because it wastes the whole day with useless dithering.

"This has had predictable effects," the expert added, "on flight controller proficiency and procedures verification."

Yet in late 2000, as the launchings grew nearer, I began to hear news of a steady improvement in the simulation and flight software. Slow progress was also being made in certifying both the workforce and the workarounds on which the success of the effort would depend. The results in flight would determine whether enough repairs had been made.

In large part, NASA's official decision to fly "with risk," in the words of one report, had been justified because of the nature of the missions. Unlike practically all of the previous NASA astronaut missions, the space station involved no life-or-death short-term urgent decisions. The space station was going to be in a stable orbit, giving engineers time to recover from computer system crashes and even hardware design flaws. Reassuringly, the U.S. segments were attached to fully functioning Russian modules, so that the ultimate down mode for any software crisis would have been to revert to Russian-only control for the hours or days (or longer) that it took to recover.

This approach had to work, space workers believed, and they expected it to work. After complaining about Russian delays for more than two years, NASA was absolutely committed to launching its three big sections on *Missions 3A, 4A,* and *5A* at the promised times. "So right now we are still very optimistic that the schedule we have in front of us is makeable," Holloway concluded at the September press conference.

"And if it's not, it's a matter of days, not months. Things are OK," Holloway promised, as he rushed from the conference to confront another last-minute crisis. Five months later, with the three missions accomplished with astonishing ease, Holloway and his team could be forgiven an "I told you so" or two. But they were probably too busy with the next series of add-on missions.

They had essentially accomplished a replay of the NASA *tours de force* of *Apollo* (1968–1969) and the space shuttle (1979–1981). In both cases, dauntingly complex technological issues had loomed over the program. As each program approached flight, a host of hitherto unrecognized problems appeared. Since each problem had to be overcome, progress toward the launches came to a halt. As the months passed and the first flights got no closer, many people worried that the projects were indefinitely stalled.

But in the end, as they say at NASA, all the ducks got into a row. For *Apollo*, the first successful orbital flight with astronauts occurred in October 1968, and the actual lunar landing took place nine months later. As for the shuttle, after spending two years marching in place at "launch −12 months," the program resumed its motion toward launch in March 1980, and launched within 13 months.

For the *International Space Station*, the year 2000 was a reprise of the triumphs of earlier NASA generations. At the beginning of the year, the first crew launch was planned for late October. The *Lab* module had been slipping for more than a year, and Holloway had been heard telling associates that he'd be happy to get it into orbit by Christmas. A year later, these two milestones were reached with no significant delay.

Ahead of the *ISS* were more crises, however, both diplomatic and financial. Technological frustrations would continue, and setbacks in the assembly process were expected. But in the success of *Expedition-1* and the *3A / 4A / 5A* "big push," America's space team shone and so did those of the other partners.

17

Last Days of *Mir*

"We easily believe that which we wish."
Pierre Corneille (1606–1684), *Le Baron*, Act I

"Truth is stranger than fiction, because fiction is obligated to stick to possibilities; truth isn't."
Mark Twain

For ten years, the Russians had been sending a new crew to their *Mir* space station every six months. They had determined that half a year is an optimal tour of duty in space. It's long enough to get your "space legs," but it's not long enough to make you "space happy." Many cosmonauts have made two or more such space expeditions.

But the 10 continuous years of human occupation in space ended in September 1999. *Mir*'s retirement was at hand, but many Russians found the idea unacceptable.

"It would be a crime against humanity to sink the orbital station," veteran cosmonaut and Kremlin space adviser Aleksandr Serebrov told news reporters in mid-September 1999. Yuri Baturin, a Kremlin official who had visited *Mir* on an inspection trip a month before, agreed: "It's a sin to deorbit such a laboratory." And veteran cosmonaut Anatoliy Solovyov told Reuters that NASA's insistence on terminating *Mir* was a plot to subjugate Russia: "It's a purely political question that there is pressure for us to get rid of *Mir* as soon as possible," he said. "It is clear why. Who has the station? We do."

This point of view rapidly gained ground. "The premature disposal of *Mir* makes as much sense as dumping your old but still working car before even buying a new one," said Aleksey Mitrofanov, chief of the Duma's geopolitical affairs committee. The Space Council of the Russian Academy of Sciences issued a report stating that *Mir* was still usable and should be kept operating for another five years. According to Boris Ostrumov, deputy general director of the Energia Corporation, which owned the Russian station, "speaking purely objectively, *Mir* could go on operating for another ten years."

Mir was a symbol of the once-proud Soviet space program. But while team after team of cosmonauts were busying themselves with experiments commissioned by various scientific institutes back on Earth, nothing useful was being transferred into the Soviet economy. In cold-blooded hindsight, other than pride, the Soviets got little good from the project.

I could understand the emotional reasons that prevented the Russians from accurately assessing *Mir*'s worth. But I also found it interesting to see how long it took NASA, blinded by its own irrational mindset, to discover the reality of *Mir*'s science research potential. The most bizarre aspect was that NASA started off close to the right answer and proceeded to get farther off course as time passed.

In 1991, congressional leaders had approached the then director of NASA, ex-astronaut Admiral Richard Truly, to examine the prospects of expanded U.S.-Soviet space cooperation. One obvious question concerned the potential scientific utility of NASA astronaut visits to the *Mir* space station. The result was a letter, dated October 23, 1991, entitled "NASA Assessment of Current Soviet *Mir* Space Station and *Soyuz* Spacecraft."

"*Mir*... is reported to be increasingly subject to the sorts of system failures and added maintenance requirements that would be anticipated in a space system originally designed for a relatively short lifetime," the report stated. The crew must "devote most of their time to housekeeping, maintenance, and physical countermeasures."

"The noise and vibration produced by *Mir*'s equipment can negatively impact the experiments," the report continued. Further, "*Mir*'s power levels, as compared to those planned for *Freedom*, are relatively low."

There were a few technical errors ("Unlike Space Station *Freedom*, *Mir*'s solar arrays are fixed"), and the report didn't always fully anticipate the value of add-on modules ("It is unlikely that *Mir* could accommodate larger, more capable modern laboratory instruments"). But overall, NASA's experts provided Admiral Truly and Congress with a fairly sound assessment of *Mir*'s utility as a science research platform.

Then, for political reasons involving U.S.-Russian relations, it became important to portray *Mir* as a wonderful opportunity for astronauts to do scientific research in space. But as plans moved forward, disturbing new evidence was uncovered.

Ulf Merbold, a German astronaut who had flown on two American *Spacelab* missions, visited *Mir* for a month in the autumn of 1994. He briefed NASA controllers on December 20, 1994, about his experiences. In notes written and distributed by NASA engineer John Graf, Merbold's comments highlighted a series of problems that NASA management would learn on its own only in later years.

"The *Mir* flies without any sort of [relay satellite], the only communication link is when over Russia," the account read, paraphrasing Merbold. "There was one relay satellite, but NPO Energia has to pay to use this satellite and they can't afford it, so they stopped using it. Many parts of the day there is no telemetry, and no communications at all. Merbold says that it is very hard to do science this way. The Principal Investigators don't know anything [about what is happening], they record data onto a diskette, and don't see any results until post-flight. Often there was a problem, and nobody detected it, and the experiment failed."

In summary, Merbold told NASA engineers that *Mir* is "no place for science." With "severe limitations for power, communications, and down weight it is very hard to do science," he stressed. This was a view that NASA didn't want to see.

NASA's workers couldn't avoid seeing it, however. During Merbold's mission, a team of American flight control experts had visited Moscow. They talked with European Space Agency controllers who were there for their mission, and they got the same story. One of them briefed me privately on his return. The Europeans had told him that their whole project was "a joke," that there wasn't enough on-board electric power to do their experiments, and that there wasn't enough telemetry or radio time to communicate anything useful with Merbold. It was all for show. This was in 1994, remember, before the U.S. visits had even begun.

Merbold had wasted days searching the station for European equipment sent up on earlier flights; some of it was never found. The main Russian "materials processing" furnace was broken, and promises to fix it before the mission had not been fulfilled. An embarrassing power failure halfway through the visit had had everybody hauling battery packs back and forth for days. Radio communication with the ground was so bad that at times the scientist had to use a ham radio rig originally on board only for recreational use. During a critical redocking maneuver in their Soyuz spacecraft, the crew had been startled by sudden darkness and loss of power even though they were on the sunlit side of the Earth, because Mission Control had forgotten there was a solar eclipse that day and that the orbit passed right through the Moon's shadow.

My friend provided more details from the mission: "Merbold said the *Mir* was very noisy, much noisier than shuttle. Many ventilators were running, and most *Mir* cosmonauts slept with earplugs. You have to negotiate a quiet place to sleep, and he considers the *Mir* to be poorly organized. A future station should have a quiet place to sleep, a noisy place for exercise, etc. The noise seemed to be Merbold's most troublesome environmental factor."

These were the same problems that would later take NASA by surprise during its own *Mir* expeditions in 1995 to 1998. Some of these problems, such as the excessive noise, were still tormenting crew members aboard the *International Space Station* in early 2001. As for how NASA managed to let itself be surprised, the old proverb says it all: "None so blind as those who will not see." Or maybe it should read, "None so deaf...."

It wouldn't really have been hard for NASA to get an honest assessment of *Mir*'s true science potential, if it had wanted to. In the late 1990s, NASA had contracted out most of its historical records work and the writing of most of its program overview reports. One of the primary historians who worked on the projects was David Portree, who later left NASA and now lives in Arizona. He had full access to all of NASA's internal documents on the *ISS* and shuttle-*Mir*. It was clear what that raw evidence showed.

"*Mir* is a terrible [science] platform," he told me in February 2000. "Don't believe NASA propaganda to the contrary," he continued. "Science was used during shuttle-*Mir* as a justification for policy, so NASA made a big deal about it to cover that fact." He specifically mentioned "poor power and poor micro-gravity" as key weaknesses.

Other working-level troops had reached the same conclusion. Bruce Cornett, a scientist in Huntsville, Alabama, built and flew the Optical Properties Monitor payload for Jerry Linenger's tour of duty on *Mir* in early 1997. "Very complex experiments require many things from the spacecraft," he explained to me in 1999. These include "power, data services, cooling, microgravity, atmospheric support, and both crew and ground interaction."

The results on *Mir* were dismaying. "Every day we ran into something else that stunned us as far as how crude accommodations were compared to shuttle-*Spacelab*," Cornett informed me. "You wouldn't believe how many power interruptions we had." For the Russians, this was nothing abnormal. "Requests for things that are routine on shuttle, like a need for a continuous supply of power with reasonable voltage regulation, blew their minds," he told me.

How about microgravity, the absence of shock, vibration, and shoves from the spacecraft? "Not too good," he replied, because of the way they used those reaction wheels to control attitude. Equipment cooling? "If the cabin air doesn't do it, it isn't there." Real-time data and ground commanding? "Nonexistent." Atmospheric composition regulation? "Not only does it not exist," Cornett stated, "there isn't even any data available on what the atmospheric conditions are, which tends to invalidate results of life science experiments."

"This has always been the problem with *Mir*," he explained. There is a lack of what NASA calls "payload accommodation." "They [the Russians] never developed a proper concept of payload operations as a distinct discipline," he concluded. "Therefore, *Mir* never had more than the most rudimentary payload accommodations."

The low level of scientific research that *Mir* could support quickly became apparent to NASA astronauts, whether or not they had ever heard of Merbold's observations from his 1994 trip. "When it became obvious that the scientific goals of Norm Thagard's flight [in 1995] could not by any stretch be achieved," a veteran doctor-astronaut told me privately, "we were told that the requirements would be changed such that whatever he achieved would become our goals." As a result, the astronaut explained, "We'd declare anything as 100 percent success."

Following the 1997 *Mir* disasters, scientific research was deemphasized even further. On July 24, mission planning official Rick Nygren e-mailed the astronauts training for the next expedition with the news that "we made *big* changes today" by deleting some tissue culture experiments and delaying others. Wendy Lawrence, who had been designated to fly the next mission, shot back an angry reply: "You all have succeeded in removing the *only* experiment that I was really looking forward to conducting, one that I actually was able to interact with," she wrote. "Good thing Frank [Culbertson] said, during Monday's press conference, that science wasn't the top priority of the 'Phase 1' program because it is no longer mine." Much of the key experiment was ultimately restored, but Lawrence never performed any of it. For other reasons, she was removed from the crew and replaced by Dr. Dave Wolf, who was able to conduct a small program of interesting science despite the difficulties. Had the difficulties been appreciated sooner, the science program might have been better designed to fit more efficiently into the limited situation.

Most of these problems were actually fixed on the *International Space Station*, but then again, they had already been fixed on the space station *Freedom* design in the late 1980s. NASA engineers already knew how to do it right; they had no need for *Mir*'s bad examples, and learned nothing useful from them.

You couldn't tell this to a Russian, though, except maybe old Konstantin Feoktistov, cosmonaut and curmudgeon, who bitterly denounced the wasteful spending on most of the Soviet showcase programs. For Russia, *Mir* was a shining city in the sky, the incarnation of the best aspects of "Soviet reality." It was practically a religious relic.

To make things worse, in the eyes of most Russians, NASA was the villain behind the plans to terminate *Mir*. The American space agency insisted that it had nothing against *Mir* itself. Space officials were only concerned that the financial resources needed for the Russian segments of the *International Space Station* might be diverted to *Mir*. At a press conference in Houston on November 13, 1998, NASA's *ISS* program director Randy Brinkley explained that the U.S. side believed that Russia simply did not have the resources to build and launch supply ships to both space stations. This time, NASA was right.

In the meantime, the bankrupt Russians were getting no benefit from the *International Space Station*, only more and more costs. They expected to have little if any research time on board, and practically no scientific equipment. Whatever cash NASA was passing over wasn't enough to cover the operational expenses of fruitlessly holding up their end of the original *ISS* bargain. From the Russian point of view, it looked more and more like a bad deal.

And Russia really had developed a strong attachment to *Mir*, the last symbol of a once-mighty stand-alone Soviet space program. NASA tolerated this affection as long as *Mir* seemed doomed, on the verge of terminal breakdown or official abandonment.

On September 29, 1999, the last crew had left the station, and for the first time in 10 years, there were no cosmonauts on board. This was officially represented as a step toward the deliberate deorbiting of the station the following spring.

Past Russian attempts to run the station under autopilot had not been encouraging. The control computers always broke down eventually, and the station lost pointing control. "Eventually" sometimes meant every week or two, although occasionally it meant within a few hours, and once within 15 minutes. Without the ability to face the station in the desired direction in space, its solar arrays would turn away from the Sun and its electric power would fail.

After each previous breakdown, cosmonauts already aboard the station had manually restarted the computers after making the necessary repairs. But now the station was to remain without cosmonauts in orbit for months, eliminating any such possibility.

The answer was a new and highly reliable analog computer called BUPO, for *blok upravleniya prichalivaniyem i orientatsiey*, or "unit for control of docking and orientation." It was hooked up to *Mir*'s autopilot in July 1999 as the last crew was preparing to leave. After testing it, the crew activated it on the day that they left the station. It would go on to operate flawlessly for another year and a half.

BUPO ran a minimum level of on-board equipment, mainly the radio link with Earth. It used *Mir*'s original small rocket thrusters to orient the station, since there wasn't enough electricity to run the efficient but power-hungry gyrodines (which turn the station by spinning motorized counterweights faster or slower). BUPO kept *Mir* in a slow spin, orienting its solar arrays more or less toward the Sun and providing adequate power for the station's hibernation mode.

When necessary, Mission Control in Moscow was able to command *Mir*'s main computer to carry out specific tasks. But for more than six months, BUPO alone successfully guided the dormant *Mir* through space.

Meanwhile, throughout most of 1999, as *Mir* cosmonauts prepared to abandon the station and then finally, reluctantly returned to Earth, there was a groundswell of Russian public opinion to somehow preserve the station. Space officials, cosmonauts, politicians, scientists, and other respected figures paraded before the Russian news media seeking to keep *Mir* going beyond the promised termination date.

At the same time, NASA was under the impression that Russia had solemnly promised to deorbit *Mir* and concentrate its space efforts on the *International Space Station*. NASA and outside experts knew that Russia's space resources were inadequate to support both programs simultaneously.

Although Russia's most obvious constraint was in building spacecraft and launch vehicles, it was also severely limited in other resources required for human space operations.

As mentioned earlier, the cadre of experts who staffed the Russian Mission Control Center near Moscow had already been thinned out by retirements, deaths, and attrition to pursue jobs with living wages. The team that ran *Mir* was to have been transferred in its entirety to support the *Service Module*.

An entirely different team from the Khrunichev space factory in south central Moscow was responsible for controlling the *Zarya* module (the *FGB*). Although its personnel were being severely strained by the need to run *Zarya* for more than a year longer than originally planned, the factory had a healthy cash flow from *Proton* rocket launches, and so could afford to hire enough specialists. Not so with the Energia group, which operated the TsUP.

Other human resources were even more scarce. At the Institute for Biomedical Problems in Moscow (IBMP), the medical experts who studied human reactions to space flight were all earning money working elsewhere. Scientists remained on the staffs of research institutes only when they or family members held paying jobs in the real economy.

In this environment of fiscal collapse, the Russian government had agreed to let Energia seek private funding to keep *Mir* going. The government retained formal ownership of the station, but the Energia Corporation was its designated operator and, the government hoped, fund-raiser.

Estimates of how much money was actually needed varied wildly, with Energia initially saying that it would need more than $200 million per year to operate *Mir*. When obtaining this much money proved totally impractical, Energia officials gradually lowered the desired amount.

At the same time, when Energia was seeking private funding for *Mir*, several embarrassing false alarms and wild goose chases attracted media attention. One British businessman who offered funding turned out to be long on promises and short on cash. Rumors of negotiations with China and with other countries turned out to be wishful thinking. Some Western commercial research teams expressed interest, but even when their offers were combined, they didn't approach the required level of funding.

In November 1999, the Duma, Russia's parliament, allocated 1.5 billion rubles (about $50 million) to keep *Mir* going. However, the legislation specified that the money come out of profits from the commission that licensed the commercial use of Russian aerospace technology. That agency's total income in 1999 was less than a million dollars, and so no actual cash was ever transferred to Energia.

Throughout 1999, a charitable fund was soliciting donations from Russian citizens to keep *Mir* going. But according to Aleksey Arnov, chairman of the fund, it raised only 485,000 rubles through February 2000. That was barely $15,000 at current exchange rates.

Then, in February 2000, a Western consortium called MirCorp, backed by two telecommunications multimillionaires, signed contracts with Russian space teams to prolong the life of *Mir* for commercial applications, and two Russian cosmonauts were launched to *Mir* on April 4 to perform some maintenance and prepare the station for a new series of visits. It was billed as the world's first "commercial manned space mission."

NASA administrator Dan Goldin complained bitterly about the reversal of Russian space plans. Reopening *Mir* was accomplished by using spacecraft that had originally been promised to the *ISS*, and Goldin called this "a major breach.... [I]t was inexcusable without consultations."

"It was not something a true partner would do," he told a congressional panel on April 6, 2000, the day that the new cosmonauts reached *Mir*. He admitted to "a high degree of frustration on our side," and he threatened Russia with "some very serious discussions at the highest level of the government." The Russians seemed unruffled by Goldin's threat to call them bad names.

"It is the director of the Energia company who proudly walked me through his plant and identified the tail numbers of two *Progress* and one *Soyuz* and thanked us for the support we gave so they could build them," Goldin elaborated. "It is the same person who without any consultation with NASA pulled those tail numbers to use to keep the *Mir* space station up."

Tail numbers designate particular vehicles under construction. It takes about two years to fully assemble a *Soyuz* spacecraft. Fortunately, because of delays, *ISS* didn't really need the vehicles, and new ones coming off the assembly line later would be able to replace them.

"I'm shocked, upset, disappointed," Goldin admitted in front of the congressional panel. "This administration is not pleased with the performance and attitude of Energia."

The members of Congress expressed more than mere disappointment and surprise. They were absolutely furious, and they ripped into Goldin for years of self-delusion and poor judgment.

"No other recent problem in our space program has cost the American people so dearly, both in money and lost opportunities," thundered Dana Rohrabacher, chairman of the House Space and Aeronautics Subcommittee. "Nothing has been so destructive as the naïve assumption that Russia's government would spend its limited resources to help us build the *International Space Station*."

"We're fools for depending on them," he added. Calling the decision to put the Russians in the "critical path" of space station assembly "a grievous mistake that has finally caught up with our space program," he concluded: "I believe the Clinton administration's decisions have put the space station and the American space program at risk."

The issue of the specific *Soyuz* vehicles to be sent to *Mir* created a serious conflict between NASA and its Russian partners. Not only had

Russia diverted the spacecraft promised to the joint project, threatening the *ISS*, but, as Goldin complained to Congress, the Russian vehicles had been finished only thanks to American money.

The *Progress* that was launched to *Mir* in February 2000 was from production line 250, the first of a new improved version called the *Progress-M1* class. It had always been allocated to *Mir*. NASA had no problems with this, especially since the Russians had always said that this vehicle would carry the fuel to deorbit the old station.

Russia also produced three human piloted *Soyuz* vehicles that year, tail numbers 204, 205, and 206. One was used to send the new crew to *Mir*. Following that launching on April 4, 2000, Russia announced that a second crew would be launched to *Mir* in September.

However, long-standing NASA plans called for two *Soyuz*es to be ready for the *ISS* in the summer and fall of 2001. One was reserved for *Expedition-1*, the first long crew to be commanded by Bill Shepherd (to be launched in October). The other was allocated to the so-called *Expedition-0*, the all-Russian crew that would stand by to launch in September in case of a docking problem between the *Service Module* and the rest of the *ISS*.

That made a total of three *Soyuz* vehicles, two for the *ISS* and one for *Mir*. Only two were available, however, and they were both largely financed with American money. And one of them was going to *Mir* with the second commercial cosmonaut expedition.

Meanwhile, three robotic *Progress* flights were also scheduled for the *ISS* in 2000. This was in addition to an already planned second supply ship, *Progress #252*, launched to *Mir* in May.

Here, too, *Mir*'s needs trumped those of the *ISS*. At a meeting in Moscow in February 2000, the Russians told their NASA counterparts that one of the three other promised *Progress*es (Nos. 242, 253, and 254) had also been reassigned to *Mir*. They insisted that Goldin had approved it, but nobody at NASA had heard of the agreement. And they expected that if *Mir* reoccupation extended further into 2001, the other *Progress* vehicles would also be needed there. They would therefore become unavailable to the *ISS*.

This was not just a problem with supplying the astronauts on *ISS* with clean underwear and science gear. The *Progress* vehicles carry rocket propellant, which is the primary means of countering the slow orbital decay of the station due to air drag. Without the long-promised propellant supplies, the station's orbit would soon decay below the required safe altitude.

NASA's space shuttles are able to use some of their excess propellant to raise orbits by small amounts, but even in a reboost-only mode, the shuttle fleet wouldn't be able to fly often enough to keep the *ISS* in orbit. NASA was building an "Interim Control Module" and a permanent "Propulsion Module" that could carry large amounts of fuel, but it couldn't afford to complete either. Both projects were canceled in late 2000.

Amidst the controversy, the commercial space mission involving *Mir* was being carried out. On January 26, 2000, the slumbering space station awoke, in response to commands from Earth. The station restarted its main computer and powered up its precision pointing devices, the gyrodines.

On February 1, the *Progress M1-1* supply drone was launched. Two days later, it docked to *Mir*. Using the rocket fuel originally allocated to deorbiting the station, the *Progress* instead pushed *Mir* even higher. Over about a week, *Mir*'s altitude was raised from 315 to 360 km, and later to 400 km (the *ISS* normally orbits at 380 km).

The computer was then returned to hibernation for another two months, until just before the launch of the *Soyuz TM-30*, carrying what the Russians called Expedition 28 to *Mir*. Cosmonauts Sergey Zalyotin and Aleksandr Kaleri reached *Mir* on April 6.

The Amsterdam-based MirCorp and Moscow's Energia Corporation formally signed a lease for *Mir* in London on February 17. While Energia owned 60 percent of MirCorp stock, Western investors vowed to invest $150 million in upgrading *Mir*'s capabilities both for research and for other commercial activities, such as advertising and tourism.

One activity that was predicted to occur was the establishment of an Internet portal on *Mir*, linked to Earth via relay satellites (yet to be launched). Color video of the Earth from space would be sold to advertising companies. But the commercial market for this was obscure, and there were loud skeptics in the space community. Sure enough, it never happened.

Andrew Eddy, who formerly headed *ISS* commercialization planning for the Canadian Space Agency, was MirCorp's senior vice president for business development. He tried to downplay the notion that *Mir* and the *ISS* were enemies or even competitors. "It's very much a complementary initiative," he told news reporters in London. "Once we start flying commercial clients on a regular basis to space, that will be very much to the benefit of the *ISS* partners." This was because, he continued, once the *ISS* becomes operational, "there will be a pool of commercial clients who are used to flying in space, who understand the value of space activity to their business."

The main financial backing for MirCorp was Walt Anderson. He had made his fortune buying and selling telecommunications companies after the breakup of AT&T. He had also invested in several other space-related projects dedicated to introducing new technologies to space operations. Anderson's investment group was called "Gold and Appel," humorously named after a fictitious futuristic organization in Robert Anton Wilson's cult science fiction classic, the Illuminati trilogy.

The project began when space activist Rick Tumlinson introduced Anderson to Jeffrey Manber, who was serving as Energia's representative in the United States. Manber, in turn, brought Anderson and Energia together. Over a period of months in late 1999, the Russians and the Americans circled each other warily, sizing up each other's credibility and capabilities. By December, their new partnership was formalized in several joint documents, and two months later, MirCorp formally leased *Mir* from Energia.

Manber used all of his diplomatic skills to smooth over relations with the Russians, with NASA, and with other U.S. government agencies. He even tried to reduce the new hostility between NASA (and its congressional supporters) and NASA's Russian partners.

For example, in his testimony before Congress early in 2000, Goldin bitterly accused Energia of "double bookkeeping" on prices for spacecraft, charging NASA one price but offering the same spacecraft to MirCorp for much less. Manber called Goldin's accusation "frustrating." He explained to me that there were good reasons for the different pricing. "The Russians are majority stockholders in MirCorp, so they will make money in three ways," he said. "From us directly, with a contract [and] when we get customers, and finally share [value] appreciation."

Despite these heroic efforts to save *Mir* and mollify NASA, one troubling fact remained: *Mir* was not (and never has been) a very good platform for sophisticated research. Nor would its orbit really allow it to service and repair other satellites. To experienced space-flight operations experts, many of MirCorp's promises seem naïve and unworkable.

On the other hand, the idea of expensive tourist flights was not considered to be too far-fetched. True, the small number of people with the assets and inclination to spend $20 million for a week in space don't usually have the spare time for the three to six months needed for training. And a potential deal with an actor in orbit was scrubbed a month before the *Soyuz* launch when the sponsoring film company failed to make the required advance payments in time. Nevertheless, Manber was upbeat at the contract signing ceremony in London: "We believe the story of the *Mir* space station and our efforts to keep it in orbit will be one of the great stories of the decade and perhaps of the century."

"The tragedy will not be if we fail," he suggested. "The tragedy would be if we didn't even try."

Try they did, through trying times. Throughout the summer and fall of 2000, the protagonists struggled with the financial issues of prolonging *Mir*. In the end, the money needed to keep it going did not materialize. Private financial backing waned as the stock market fell, and liquid assets for further investment became scarce. NASA officials and pro-NASA congressional staffers waged their own underground campaign against potential U.S. clients for MirCorp, explicitly threatening to withhold *ISS* privileges from anyone who paid for a *Mir* mission. (NASA had a habit of stabbing its competitors in the back; for example, it had successfully lobbied against a much cheaper private alternative to the *Freedom* space station in the late 1980s.)

But the money wasn't the real issue. Once the *ISS* really became inhabited in November 2000, *Mir* was doomed. This was because the Russians were producing only enough *Soyuz* and *Progress* vehicles to support one station. Since it takes 18 to 24 months to build each of these vehicles from scratch, even the appearance of unlimited funding in mid-2000 couldn't have saved *Mir* unless the Russians had deliberately chosen to abandon the *ISS*. They couldn't have built twice as many spacecraft in the time required.

NASA had been studying alternative *ISS* strategies based on the loss or further delay of Russian contributions. Russia, it turned out, had its own contingency plans for such failures. Their backup plans centered on *Mir*. It made good sense not to terminate the station while there was still a chance that they wouldn't be playing a big role on the *ISS*. But once the *ISS* got going in earnest, this alternative was too expensive to continue. *Mir* had to die.

It wasn't entirely the fault of the outside world. *Mir* carried the seeds of its own doom internally, as well. Revived after the 1997 crises by many shuttle-loads of new equipment, the station then experienced the normal wear and tear of space flight as equipment aged or failed. The new set of electric power storage batteries was degrading. A pesky slow air leak that had begun in mid-1999 worried engineers and cosmonauts. The normal delivery of replacement and repair equipment had practically ceased, and more and more of the critical equipment was operating far beyond the original lifetime guaranties. The chances of a new catastrophe rose daily.

MirCorp reinvented itself as the commercial arm of RSC-Energia and looked forward to activities aboard the *ISS*. It had kept faith with the Russians, a major factor in future dealings, and it had gained a more realistic idea about what people would be willing to pay for in space.

In mid-January 2001, the *Progress M1-5* robot was launched to *Mir*, with a special cargo section containing extra propellant tanks. There was also a supply of food and water on board in case an emergency crew had to be sent up to restore Earth's control over the station. A two-man crew was trained and ready, and the *Soyuz-206* vehicle was standing by on two weeks' notice at Baykonur.

The plan was to let *Mir's* orbit gradually decay under the influence of atmospheric drag. During these months, the drag was much higher than normal because of the peak of the Sun's 11-year activity cycle, which energized Earth's upper atmosphere like a bowl of popcorn. Molecules bounced much farther into space than they usually did, and there they were hit by the speeding *Mir*. Although the retarding force was only equivalent to a finger push (1 to 2 newtons in metric), it was continuous, and toward the end, *Mir* was losing about two miles a day in altitude.

Normally, the engines of docked *Soyuz* and *Progress* vehicles were used to nudge the station higher. But this time, the *Progress* would use its engines to slow the station down enough for it to slip into the upper atmosphere and be destroyed.

But this maneuver required delicate timing. The problem was that even if the propellant tanks were burned to depletion, the maneuver would be able to slow the entire complex by only about 200 ft/s from its total orbital speed of roughly 25,000 ft/s. This was enough to lower one side of its orbit by about 70 miles.

To guarantee that the station would be "captured" by atmospheric drag and wouldn't slip back out into space, the orbit had to end up no more than 60 miles high. This meant that the station, which normally circled at an altitude of 220 miles or higher, had to be allowed to descend naturally to an orbit about 130 miles high. Only at that point would the rocket firing have enough push to reliably "sink" the station.

But it was risky to slip that low. First, the thin atmosphere exerted twisting forces on the vanelike solar panels, forcing the station to use precious propellant to control its pointing direction. Second, this took time, and every hour was another opportunity for something else to go wrong. And finally, the lower orbit meant that the station didn't rise as high or stay up as long when it passed near Russian tracking sites, and this meant that the already too-brief radio contact periods were even shorter in duration and less frequent.

For safety's sake, the rocket burns would be performed only during the 15- to 20-minute-long intervals when the station was in direct contact with Russia. Because the *Progress* engine was so small (about 660 pounds force) and *Mir* was so heavy (130 metric tons) it would take more than an hour of rocket firing to achieve the maximum braking. So that total had to be broken into small segments. This was also advantageous because the thrusting was effective only if it was within five or ten minutes of the high point of the orbit.

Scary stories about flaming space debris had elevated anxieties around the world for weeks. *Mir* was the biggest manmade object ever deliberately crashed onto Earth, but over the years there had been a number of others nearly as big that had received little attention. There had been *Skylab*, *Skylab*'s *S-2* booster stage, the 1987 *Polyus* space battle station and its booster, the two-module *Salyut-7*, and a number of earlier *Salyut*s. Then there had been the up-to-20-ton U.S. and Soviet satellites that were routinely dumped into the Pacific, successfully every time. On each of its hundred launches over the last 20 years, NASA's space shuttles dropped 40-ton fuel tanks into the Indian and Pacific Oceans at nearly "orbital velocity." During 40 years of space flight, nobody had ever been hurt. And that didn't even count the natural objects of *Mir*'s size and larger, since 5 or 10 of them randomly fall from deep space every year and at much higher speeds.

There was a lot of screaming and shouting by nationalist and communist politicians in Moscow, and a number of veteran cosmonauts (Vitaliy Sevastyanov and Svetlana Savitskaya in the lead) demanded that *Mir* be saved. Even the last cosmonaut on *Mir*, Aleksandr Kaleri, didn't like the idea of destroying it, despite his new flight assignment aboard the *ISS*. "To have one's own space program is one of the characteristics of a strong state," Kaleri noted. "Only a few countries can afford themselves this luxury. While we still have *Mir*, we are still a great power. But once we lose it, we will slip away from the club of great nations."

But Kaleri's crewmate on the last mission, Sergey Zalyotin, told news reporters that merely visiting *Mir* occasionally was too much trouble: "It takes four or five weeks to open it up and make repairs before you can begin work." Sergey Avdeyev, who spent a total of more than two years aboard *Mir* on three expeditions, agreed that the station's condition was deteriorating: "Future missions would only become more and more difficult," he told an American journalist. And Musa Manarov, the first man to spend a full year on the station (1987–1988), also wasn't nostalgic: "It is impossible to create something new if you keep clinging to the past," he noted.

When it came time for *Mir* to die, nothing went wrong and nobody got hurt, as usual. On March 22–23, 2001, controllers in Moscow commanded the station to perform three sequential braking maneuvers, each one lowering the station's orbit, and the last one placing the station on a suicide trajectory into the empty South Pacific. Or at least it was thought to be empty until two dozen American tuna fishing boats called in. They weighed the odds and opted to keep fishing.

As part of CNN's coverage of the station's death, I got to play "the last man on *Mir*." The program was hosted by correspondent Miles O'Brien in Atlanta, with *Mir* astronaut Norm Thagard and other specialists around the country as guests. I was stationed in a *Mir* mockup in a museum in Wisconsin. We ran live broadcasts for U.S. programming, for CNN's international service, and for a special Webcast.

My job was to make *Mir* seem real. I showed the kinds of hardware (storage tanks, mostly) that would be likely to survive entry and be found floating in the ocean. I even speculated about what it would be like for somebody on board *Mir* during the final plunge. But I didn't forget to remind Miles O'Brien, several times, that he'd have to beam me off the falling spacecraft in time.

Then we heard from CNN's correspondent on Fiji, who had seen the fiery debris pass overhead. He described half a dozen flaming fireballs, with a few smaller ones breaking off. And he even caught it on videotape.

Just before the final seconds when *Mir* was peeled open like a banana and burned into a blizzard of confetti, I had my final broadcast as the make-believe last man on *Mir*. "Now would be a good time, Mr. O'Brien," I wisecracked, reminding him of his *Star Trek* promise (after all, "Miles O'Brien" had been the name of the "transporter officer" on television's *Starship Enterprise* years before). And *Mir* ceased to exist.

18

Day-by-Day
in Orbit

"Learn what is true in order to do what is right."
Thomas Huxley

*"Adversity has the effect of eliciting talents, which in
prosperous circumstances would have lain dormant."*
Horace

L ate in the evening of January 4, 2001, *International Space Station*
commander Bill Shepherd was completing his daily entry in the
"Ship's Log," his written account of the crew's accomplishments, frustrations, and observations. In describing the crew's after-dinner relaxation,
he noted that the men had viewed the first DVD disk of *2010*, Arthur C.
Clarke's story of a U.S.-Russian joint mission to Jupiter to reactivate HAL,
the homicidal computer autopilot of *2001—A Space Odyssey.*

Wrote Shepherd in his Ship's Log, "We note that the movie opens
with a recounting of 'Ship's Log' from the previous mission." The following evening, he wrote that the crew had "finished the second [DVD]
disk," and he added: "Something strange about watching a movie about
a space expedition when you're actually on a space expedition."

Long before the mission began, Shepherd had been determined
to imprint U.S. naval tradition on the space station. He found the psychological parallels between sea voyages and space station expedi-

tions to be very compelling, and his own military background was in the U.S. Navy.

We talked about it one afternoon, about a year before the launch. I reminded him that the first crew aboard the *Skylab* space station in 1973 had also been all Navy. That crew, too, had used nautical rather than aviation terminology for equipment, directions, and structural features—"bulkhead," "galley," "head," and so on. Shepherd quickly chased down Dr. Joe Kerwin, one of the members of that crew, and they began to discuss more of the naval traditions.

One common naval tradition is a written Ship's Log, kept by the captain of the vessel. Shepherd decided that he would keep one. These daily one- to two-page notes were stored on a NASA computer in Houston. In the beginning, the notes were freely available to the public. Then, two months into the mission, NASA managers decided that their contents might be private and proceeded to restrict access to all of them—in spite of the fact that Shepherd had explicitly told reporters that he was writing the notes for the public.

"We are going through the process of determining if these logs are Shep's private information," explained space station program manager Tommy Holloway. He said that he and Johnson Space Center director George Abbey would decide "how much of it should be distributed."

Fortunately, I already had archival copies of the reports that had been made public. They were full of day-to-day details, catalogs of frustrations and triumphs, and candid advice to Earth about how things could be run better next time. And behind the space jargon were Shepherd's sharp eyes and dry wit.

The crew's descriptions and comments are the moments of truth, the culmination of the years of blood, toil, tears, and sweat for the American, Russian, and international workers on the *ISS*. What worked, and what didn't work? What was expected, and what caught them by surprise? The taste of space is in the words of these men. This isn't speculation or supposition; it isn't make believe. It really happened. And it never happened quite this way before.

Early in the flight, the crew was treated to an unearthly view. It seemed to signify that they had really passed beyond the realm of the known world. "November 10, 2000. 11:30 GMT," the entry stated. "Transited through a very unusual aurora field. Started as a faint green cloud on the horizon, which grew stronger as we approached. Aurora filled our viewfield from *Service Module* nadir [down-facing] ports as we flew through it. A faint reddish plasma layer was above the green field and topped out higher than our orbital altitude." They definitely weren't in Kansas any more.

They were lucky to have spotted the natural wonder, since they could average no more than about 10 minutes of sightseeing per week. The mission kept them incredibly busy. But they occasionally found time to notice the outside. On January 3, Shepherd wrote: "Up at the usual 'crack of dawn.' We are orbiting where we get continuous light now, so 'dawn' doesn't really apply to us. Kind of like being above the Arctic Circle with the Sun right on the horizon." It was a down-to-earth description of an unearthly sight.

Thanks to the Russians' experience on long space flights and Shepherd's experience on long naval voyages, the crew quickly adapted to the different tempo of space station operations. They knew that their work would help future operations in the months and years ahead. But the long-term significance of their tasks was often overshadowed by the quick fixes of the moment.

Once, a supply flight that would have carried up a work table was canceled. On November 27, the log described how the crew overcame the problem of the missing table with creative ingenuity.

"Breakfast in the 'wardroom' on our mess table," Shepherd boasted. He explained what he and his crewmates were so proud of: "The crew 'turned to' over the weekend and we have a prototype table set up with brackets and clamps to evaluate size and location. All hands agree that the standard table (still on the ground) will not be very serviceable, as it will block access to the [freezer], and will be in the way of the treadmill and the doorway of the starboard [bunk]. Our do-it-yourself model is somewhat narrower, shorter and lower. We think our prototype will fit better, and we feel pretty happy that we can leave the other table on the ground. Power tools are out and development activity on the prototype continues as we find time."

A great many of the crew's tasks involved on-the-fly repairs. As Shepherd noted in his postflight debriefings the following March, the on-board tool kit was critical to getting the station operational and keeping it that way. This was a reprise of a familiar theme from *Skylab* and

Mir, but as usual, *ISS* planners seemed to have overlooked the lessons. "I really think we need the capability onboard to fix things while we wait on replacements from the ground," Shepherd stressed. His favorite tool was his Navy SEAL knife, which "was used every hour of every day for the first month."

The crew's desire to perform repairs in space clearly far outstripped the ground's plans to equip them with the necessary tools. For example, the crew had repeatedly asked that a coaxial cable repair kit be sent up, but the ground never accepted their justifications for it and refused. The crew also wanted a battery tester, but the ground told them to use another tool that was already on board, even though they had repeatedly told the ground that it didn't work. There were still some hard lessons to be learned in this regard.

Another major problem was the inadequate level of noise suppression aboard the modules, which forced the crew to wear earplugs for much of each day. This was yet another of the recurring problems that was never really solved. On November 24, Shepherd wrote about the sound levels: "Shep's perception of noise—very similar to [naval] shipboard environment. Noise is a distraction, but bearable. We are getting reasonable sleep, all hands wearing earplugs. What would be very useful would be to have the noise canceling headsets adapted to the comm system, as we are on a radio almost every hour during the work day."

Early in 2001, a NASA safety meeting in Houston raised the issue of noise levels aboard the *ISS*. For years, it had been known that the equipment noise inside the Russian *Service Module* was going to be too high. When the Russians made it clear that they weren't going to do anything significant about it, NASA issued a temporary waiver—called a Non-Compliance Report, or NCR—for the first several flights. This was supposed to give the Russians time to develop noise baffles for the noisiest equipment. But the Russians hadn't gotten around to it, and they needed an extension on their original waiver.

Long before launch, NASA experts had known that the noise would be a problem. A small NASA team, including some astronauts, met with Russian medical scientists in Moscow on August 17–20, 1999, to discuss the "setting of mutually acceptable standards." Note the wording. It wasn't a matter of setting "safe" standards, it was a matter of what standards the Russian side would "accept." The NASA astronaut who wrote the report on the meeting noted that "it is clear that the noise levels on SM will be very high, comparable to those on Mir," but the only suggestion for improvement that he mentioned was that "a paper is being prepared for management on this problem with proposed countermeasures."

Little was actually accomplished, despite the two years of delay in getting the *SM* launched. So Shepherd's crew was launched anyway. They made no secret of their concerns over the noise levels, and in their in-house postflight debriefings in March 2001 (which NASA kept secret), they got very specific in their criticisms.

When asked to rate the importance of decreasing the noise level in the *SM* on a scale of 1 to 10, station commander Bill Shepherd gave it a 6, adding that it was "just a tad worse than 'what we have is livable.'" He went on: "We need a strategy to fix the noise levels. We need to target the big noisemakers specifically. These were the thermal system pumps, the air conditioner compressor, and the Vozdukh valves." The noise, according to Shepherd, "interfered with communications," although the crew could still hear the alarm tones. "Unpleasant" was how Shepherd rated the noise, but he said it was "fixable."

"Vozdukh affected our sleep," *Expedition-1* cosmonaut Sergey Krikalyov told a postflight debriefing meeting. "The worst thing about it is that it is not continuous. Every 10 minutes there is a loud noise. In the *Mir* it was located in a different module. I always said having it in the *SM* compartment was not a good idea." But that's where the designers, who apparently weren't interested in the experience of actual crew members, put it anyway.

Yuriy Usachyov, Jim Voss, and Susan Helms made up the crew for the second expedition, which took over in March 2001. On the issue of noise, they agreed with their predecessors. In a progress report sent back to Houston, but not released to the public, they noted that "noise is still a problem, with the SM being the noisiest area (68 to 70 db).... If sound suppression materials were made available, crews could continue to reduce sound levels by insulating noise makers."

Trash stowage was another aspect of the massive inventory management problem on the space station. Experience from *Skylab* and *Mir* had showed that this was going to be a problem. Stowage of and access to equipment and supplies had been a daunting challenge on both stations.

But the contract team that had taken care of the problem on shuttle-*Mir* in 1995 through 1998 had been laid off at the end of the program (the manager got a job supervising baggage handling at a Houston airport). The novice specialists on *ISS* had to learn the same things all over again. What's more, *Expedition-1*'s logs were full of descriptions of a nonfunctional automated stowage bookkeeping system, critical descriptions that were excised from the versions shown to the public.

Proper space management was more than merely a matter of convenience. Regarding the struggle with stowage, Shepherd's postflight comments were merciless. There was so many items stuffed into the *FGB*, he said, "the condition when we arrived was not livable, and I would say unsafe."

The crew faced challenges from Earth as well. In a passage subsequently deleted by NASA (probably because it might have led to hurt feelings in Russia), Shepherd criticized his support on the ground: "Another thing that would make IMS [the computerized inventory management system] much better for us would be for both Houston and Moscow to try and make the IMS world we see more of a unified 'environment.' We are getting frequent words from both sides that 'that's a Houston problem,' or 'it's up to Moscow to do that.' There are no spectators here—we are all on the team on this one."

To Shepherd, many of these struggles took on an almost military flavor. On December 13, he noted how tough the day had been, but he added historical perspective to the crew's determination to get their tasks done right: "Kind of a frustrating day, but enjoying the aspect of having plenty of time to fix this stuff later," he wrote. "Reminded of the Civil War story at Pittsburg Landing. Union troops start the battle with heavy losses. Sherman tells Grant that they had had the 'devil's own day.' Grant replies: 'Yep.... lick 'em tomorrow, though.'" After the flight, his number one observation was that "We were more busy on orbit than I think any of us anticipated."

After days of such challenging tasks, the crew looked forward to weekends. On December 16, the log notes: "0710 AM. Up and about. Saturday morning. Fortunately, a 'rest' day. We are ready for some time off. Looking for coffee." They managed to grab relaxation where and when they could, a lesson Shepherd brought to *ISS* from his pre-NASA military experience.

The holidays brought no rest from the challenging tasks, however. The crew did take December 25 off, wrote Shepherd: "0700 AM. Christmas morning. Stockings are hung on the [freezer]. Yuri and Sergei surprised with this tradition, but hey, that's what stockings are for." But the very next day, the crew manually guided the redocking of a *Progress* supply craft. The disastrous collision after the botched redocking experiment in June 1997 was on everybody's mind, but the procedures now were much safer. The *ISS* crew was also better trained and more rested than the hapless *Mir* cosmonauts.

The redocking succeeded. "Yuri gets a 'pyaterka' [the highest grade given in Russian schools, the equivalent of "A" or "excellent"] for nailing the docking cone. We opened the *FGB* hatch and the skid mark on the inside of the cone is about 7 cm. off of dead center. Opened the *Progress* hatch. Pretty cool inside—maybe 8 deg C. We are glad to have some more volume onboard for stowage."

Although the docking worked perfectly, the crew found that many of the station's systems weren't working right, including the system for reporting what wasn't working right. "We have tried several times to get the 'crew squawk' tool running," the entry for January 5 stated [a squawk is a complaint]. " We are able to log in, but the program either locks up or won't launch when we try to run it. We would like to start documenting anomalies or things which need specific tracking and we believe the squawk tool will be a good way to do this if we can ever get it to behave. We would like to 'squawk' the crew squawk for starters."

This innocuous and somewhat humorous complaint was one of the items that NASA decided to censor from the publicly released versions. "Certain operational, debriefing material has been edited from the *Expedition One* ship's log," explained NASA's Web site. "This material is considered an integral and critically important element of the on-going, deliberative decisional process NASA is undertaking related to long-duration International Space Station missions.

"This process must include necessary give-and-take communications about all aspects of crew and station performance," the explanation continued. "To be effective, these communications require absolute candor in discussion that would not be available if parties to the exchange, including intended recipients on the ground and future crewmembers, thought the material might be released to the public."

In principle, this is a valid point. The previous U.S. human space missions had been short, lasting only up to two weeks, and crews could save their most serious critical comments for blunt discussions during postflight debriefings. The transcripts from such meetings are not made public. While this may work for comparatively short missions, it doesn't work for those that last upward of four months. For long-duration missions, in-flight debriefing-style candor is a reasonable expectation, at least at certain times.

The uncensored logbook entries that I was able to review, however, suggested that NASA's deletion of large amounts of text from the officially released versions also served to limit public awareness of the difficulties and frustrations encountered by the astronauts and cosmonauts on board. A comparison of the original entries with those released by NASA shows that the deleted material dealt with specific hardware and documentation problems. The crew described what wasn't working, why it wasn't working, and what should be done to fix it.

In one deleted passage, the astronauts described a software problem familiar to millions of computer users back on Earth. "Sergei and Shep both experiencing problems with print jobs," the original entry reported. "Shep is continually getting half-page prints on some of his stuff. We are both getting blocked from changing the printer job cue. Apparently we don't have the right permissions. If the [Mission Control] folks could help here, this would be greatly appreciated." Why NASA wanted to hide this complaint from the public is baffling.

Here's another highly technical crew complaint that somebody at NASA decided was too hot for the public: "Shep tried to log all the IRED [Integrated Resistive Exercise Device] data on the computer," the deleted entry reads. "Opened the exercise and IRED applications. Went to the beginning of the IRED workout files and tried to start logging data from 14 Dec, which was when we really got going on the device. The program would not open on any file other than today's, despite what the user buttons or help file said. Shep decided that he would concede this latest round to the [computer] and go do something else. Finally sent the data down on a mail file to the Flight Surgeon. Suggested to the Flt Docs that we stay in this mode for now. Score now is about [medical computer] 8, Shep 2." Shepherd may have lost hours of work on this problem, but he hadn't lost his sense of humor.

Shepherd's written comments provided a view of space station life that is refreshingly free of spin, to the benefit of both space workers and the general public. But the tradition that he tried to establish was unable to take root. The commander of the next expedition, Yuriy Usachyov, stopped writing log entries entirely—as far as the public was allowed to see. NASA claimed that this was "crew choice" and that it was beyond the authority of Mission Control to alter.

NASA itself altered other public-insight policies. During the course of the first expedition, which ended with the crew's return to Earth aboard a visiting space shuttle in mid-March 2001, radio communications with Earth were regularly fed out live over the NASA TV channel. Conversations between astronauts, cosmonauts, and operators in control centers in Houston and Moscow were available nationwide via many cable channels or through a backyard satellite dish.

That stopped the day the crew landed. Although the voice signals are still played during working hours in the press offices of several NASA centers around the country, the public's access to them has been cut off. Rob Navias, a public affairs official at the Johnson Space Center, told me that NASA didn't any longer "have the resources to support distribution on NASA TV."

Sending the mission audio channel to NASA's Web site in Maryland for audio streaming over the Internet is an obvious technical solution, Navias admitted, but it "requires a fairly costly modification" to equipment. He refused to estimate the projected cost of such a modification. Television rebroadcast of air-to-ground conversations during Expedition-1 was conducted at night and on weekends, when other programming was not needed, Navias explained. "It required resources at JSC and at Goddard," he pointed out, and the programming has since been canceled. NASA did continue a one-hour daily summary of *ISS* operations.

Once the station's Ku-band communication system began working in March 2001, four channels of multiplexed video were coming down to control centers and payload operations teams via relay satellites. According to Navias, "by policy of NASA headquarters and the 'NASA TV' executive producer, there is too much other video to dedicate 'NASA TV' to ISS video." Instead, a weekly video highlights summary would be released every Wednesday. Navias stressed that "real time" video and audio remained available to the news media agents who visit press centers or have their own trailers on NASA sites. Since the newsroom is open less than half the time, however, all print journalists and practically all broadcast journalists were excluded.

The press-center-only policy had been followed for NASA's previous space station, *Skylab*, in 1973–1974. But back then, all space-to-ground conversations were transcribed into hard copy for the use of space scientists, the news media, and the general public. These written records provided reliable, unexpurgated insights into space station events.

But no more. NASA terminated the manual transcription efforts after the third space shuttle mission, in 1982. Despite two decades of technological progress in computer-aided systems, NASA appears to have no interest in resuming the process. Experts in the field of automated voice transcription estimate that a workable system for NASA could be operated for about $100,000 a year. Apparently, in a project costing a hundred thousand times this amount, the extra expense is unwarranted. NASA officially denied it was transcribing any air-to-ground communications, which baffled me because from time to time my friends at NASA e-mailed me transcripts of exactly such conversations.

While there wasn't at first any reason to suspect that this looming space blackout was motivated by a desire to deliberately cover up or distort events on NASA's space station, occasional lapses in candor by NASA press officials in the recent past raised the concern that a monopolized information flow would be a slanted information flow. Whether the issue was a flubbed debris-dodging maneuver in 1998, or some nonfunctional Russian space suits in 1999, or a false fire alarm compounded by a computerized checklist crash in 2000, the people that NASA hired to inform the public just weren't fully up to the task. "Happy talk" is easy, but rigorous candor about problems takes a level of effort—and a mindset—that has sometimes been lacking. The way I see it, the public's right to know shouldn't be held hostage to perpetual budgetary issues, squeamish censors, or bureaucratic inertia. But that's NASA's new openness (or lack thereof) policy.

When *Expedition-1*'s crew returned to Earth, they were candid in public about the problems that they had faced. And, as was proper, they were even more pointed in their private criticisms. Their comments also showed how much had been learned from the first *ISS* mission with crew aboard. However, most of the lessons of *ISS* should have already been learned during the *Mir* missions. NASA had spent years bragging about how much it had learned. Harsh flight experience showed that it hadn't learned much.

One critical safety element was the computer-driven alarm system, called "caution and warning" in NASA jargon. It was supposed to alert the crew to unusual situations, which could range in seriousness from merely bothersome to outright life-threatening. Yet the crew complained that the caution and warning system kept doing strange things. "It's the darkest hole we have right now," Shepherd remarked, adding that "things happened during flight that we didn't understand." Krikalyov gave an example: "We would sometimes get the 'OTHER' alarm and have no idea why." It's not good to have your spacecraft computer giving you a warning that you can't interpret.

Among the items that needed fixing was a feature of the alarm system that would annunciate for five seconds, then turn itself off. This wasn't a problem in itself, but what happened in practice was that the first alarm would mask any subsequent alarm, as no alarm would sound until the previous alarm had been acknowledged by a keystroke on the control computer console. On the other hand, if the condition that had caused the alarm ceased before the crew got to the console, the entry vanished entirely from the screen so that the crew couldn't tell what it had been. "We hated vanishing alarms," Shepherd told the debriefers.

The yellow "caution" light labeled "OTHER," said Shepherd, "was on for 98 percent of the flight," a big problem that led the crew to simply ignore it. The alarm reference books were far too big, the crew said, "almost unusable." Shepherd was frustrated that the crew had "never had a reasonable integrated caution and warning simulation during training." But they had practice in flight: For example, a false "rapid depressurization" alarm sounded so often that the crew learned to check a mechanical pressure gauge for signs of a real disaster. They soon disregarded the software system entirely.

All of these software problems should have been caught and fixed before launch, since in hindsight, it seems that none of them was unpredictable. Instead, the crew had to discover them and learn to tolerate them.

Overall, both in public and in private, Shepherd made the case that everybody had learned a tremendous amount from working on the first expedition. In large part, this was due to the candid comments of the crew. "It might have been painful at first," he told a press conference on March 30, "but in the end I think the result was good."

Shepherd, the rough-and-tough former Navy SEAL, turned out to be a first-rate space station commander. Years earlier, he had flown four good shuttle missions, after which he survived several years as program manager on the station project. As the commander of the first crew, he managed time on orbit efficiently. He knew when to order breaks, and he knew when to keep everyone up past midnight on special projects.

Initially, the Russian side had been so dubious about Shepherd that one veteran cosmonaut, Anatoliy Solovyov, had quit rather than serve under him. But even from the Russians, the verdict was unambiguous. "Our cosmonauts speak highly of human relations and work atmosphere aboard the station," Dr. Yuri Katayev, the Russian crew doctor, told news reporters after the two Russians had returned to Moscow. "They praise commander William Shepherd for his ability to establish the right tone in dealing with his colleagues and the ground services during the expedition." Considering the sources, that's high praise indeed.

The crew had high ambitions, too. Back on January 6, just halfway through their mission, Shepherd, Krikalyov, and Gidzenko were relaxing. They cast their eyes on more distant targets: "Finished the day with more email and watched *The Rock*," the log stated. "Quote of the day (from us)—'Put some more engine on this thing and send up that Mars vector.'" They had been following the same orbit around Earth for months, and apparently they were getting bored with where it was taking them. As they soon discovered, however, the boredom wouldn't last.

Ahead of the *ISS* was the cascading computer collapse of March 2001, the worst crisis in more than a decade of U.S. human space flight. At the height of the crisis, some sources told me, the station was only one hardware failure away from having to be abandoned, possibly forever. But the crew and the ground controllers held on, worked the problems, and clawed their way back to a fully operational station. It was the "baptism of fire" that many experienced engineers had expected to come along sometime. Most of them had figured that the station would survive if the teams were up to traditional NASA standards. They were.

By mid-2001, once NASA had wrapped day-to-day *ISS* communications in a shroud of secrecy, the temptation to exploit this secrecy to project a false public image became too great. On June 28, 2001, chief *ISS* flight director John Curry sent out a memo stating, "I want every *ISS* Flight Director to understand the U.S./Russian relationship issue is becoming a problem onboard." And, to keep the public and the Congress from knowing how bad things were, he added a stern warning: "Please DO NOT FORWARD!!" Somehow or other, a copy of the memo reached me anyway.

Curry's memo was sparked by a message from *Expedition-2* crew member Jim Voss, then in orbit with Susan Helms and the station commander, Yuriy Usachyov. The crew got along fine together. But on June 27, Voss had reported that the TsUP (the Russian control center in Moscow) insisted on talking only with the Russian on the station, and refused to give Voss data intended for the Russian. "We are becoming more segmented every day and the unwillingness of some people in the TsUp to deal with any crew member is making this worse," Voss wrote.

Usachyov asked Moscow for an explanation. According to Voss, an official "said that we still have no agreement that Houston is the lead control center." Months earlier, officials had told the public that Houston had assumed that role, following the activation of the U.S. *Lab.*

"So, who is the lead center?" asked Voss. "I realize that in many ways the Russians are still flying the *Mir* and we can't get them to move ahead, but we have to keep plugging away."

"As a professional astronaut with a lot of experience, I was insulted by their approach to this," he continued. "To me it is totally unacceptable for them to not work with any crew member on any subject. I hope this is resolved after Yury's discussion with the supervisor today, but I wanted you to know what we were thinking about this and what was going on off the loops. I will be patient and continue to work with them, but I will find it hard to accept another occurrence of this type."

Voss requested that this problem be concealed from the public. "This is private/personal mail and not for release to the media." This was in violation of written NASA policies on operational versus personal email.

Curry replied to Voss by another email message never intended for the public to see." I also had a very similar week regarding Russian relations," he wrote. "The Russians ARE DEFINITELY trying to force us into segmented operations," where the Russian crew members obey the Moscow TsUP and the Americans are controlled from Houston.

This is a direct violation of the fundamental principle of the "unified crew," agreed to by all the international partners.

"They REFUSE to acknowledge any type of U.S. leadership in the planning world on items such as sleep cycles, priorities, etc., and they proactively maintain a written record of instances when [Houston] makes a planning mistake. This record is then periodically shipped to Houston as justification for why they should still lead planning (i.e., [Houston] planners are a bunch of idiots who don't know their ass from a hole in the ground)."

Curry had found out that a month earlier, TsUP boss Vladimir Solovyov instructed all his flight directors to follow the policy that the TsUP was still in charge, and they should "maintain ownership of all Russian Segment related activities." Concluded Curry, "I believe Solovyov's goal is to drive us into segmented operations so they can have free reign to profit from *Space Station* operations without U.S. interference."

"I've been trying to fight this 'who's in charge' battle behind the scenes for quite some time," Curry wrote. "We have jointly signed Flight Rules which clearly state [Houston] is in charge of all operations including Planning except in specific instances such as Russian Vehicle docking/undockings and [spacewalks] from the SM," he explained. "In addition, we have signed protocol agreements defining the requirements which [Houston] must meet before complete handover can take place. [Houston] has now met all of these requirements."

In public, these conflicts weren't mentioned, and NASA denied any Houston-Moscow friction. Internally, officials knew they were covering up to ensure a public image of diplomatic concord.

The *ISS* was on a roll, with the impressive achievements of 2000 behind it, the March 2001 computer crisis overcome, and a major new level of on-orbit capability stretching ahead with near-infinite promise. The political and diplomatic foundation of the 20-nation partnership had taken some strong hits, but it had survived, and project workers saw every reason to expect this to continue. Their confidence was at least as high as that of the Soviet space team that had found itself in the same situation, with the same confidence, in 1985, where we first began this story. Back then, the Soviets soon realized, as today's space teams may yet rediscover, that although space is a vacuum, it is full of surprises. And projecting current success into the future is tricky, because orbits through space never follow a straight line.

19
Future Orbits

"We should be careful to get out of an experience only the wisdom that is in it—and stop there; lest we be like the cat that sits down on a hot stove-lid. She will never sit on a hot stove-lid again—and that is well; but also she will never sit down on a cold one anymore."

Mark Twain

"You will find that the truth is often unpopular and the contest between agreeable fancy and disagreeable fact is unequal. For, in the vernacular, we Americans are suckers for good news."

Adlai Stevenson, commencement speech at Michigan State, June 8, 1958

Even after dealing with the Russians for 10 years, NASA managed to show just how little it had learned and just how badly it could still blunder. In March 2001, the agency picked a fight with Russia that never should have been fought, a fight that could never be won. And when NASA inevitably lost the bitter confrontation, the result was anger and dismay on all sides.

The question was whether millionaire "space tourist" Dennis Tito would be allowed on board when he arrived at the *International Space*

Station as part of the Russian crew of a replacement *Soyuz* bail-out space-craft. One *Soyuz* always remains attached to the *ISS* for emergency evac-uation of the expedition crew, and since each *Soyuz* lasts only six to seven months, it needs to be replaced periodically. The Russians had always made it clear that they alone had the authority to choose the crew of the *Soyuz*, and they had made sure that all of the international documents contained the appropriate wording.

NASA knew that there would always be two Russian crew members; who they were was immaterial. The third seat, it knew, would probably be sold to a visitor from another partner in the space project. Russia had been getting $15 million or more for the third seat on visits to *Mir* in the 1990s, and since there was still little usable research equipment on the *ISS* to make a scientific visit worthwhile, nobody had been willing to meet Russia's price. Then Tito, who had bought a ticket on a flight to *Mir* before it was terminated, offered to pay for a trip to the *ISS*. The Russians signed a deal.

Cries of outrage arose from space program officials in different nations (some of whom had hoped that their own astronauts could have had the seat at a lower price). Tito's presence would interfere with crew activities, it was claimed. He would be dangerous because he would not be fully trained. And he would use the resources that space shuttles had brought to the *ISS* at great expense.

In public, NASA administrator Dan Goldin proclaimed that "safety is our number one objective," and that that was the reason why Tito couldn't go. The Russians hung tough, since without Tito's money, they couldn't afford the launch rocket. Without the rocket, the mission would have to be canceled, and the *ISS* would then have to be evacuated. It didn't help that NASA chose to prohibit Tito's launch during the week that *Mir* died, a week of mourning for the loss of the country's space independence and its subjugation to American leadership. NASA couldn't have chosen a worse time, and a surge of Russian nation-alism began to build in support of Tito and his flight.

William Readdy, a former astronaut who was serving on Goldin's staff in Washington, explained why space flight was still too dangerous for nonprofessionals. On national television, he cited the February 1997 fire aboard *Mir* and the June 1997 collision with a *Progress* freighter. These were examples of situations in which the skills of merely amateur astronauts would prove inadequate.

But these examples would have proved just the opposite, if the full stories had been disclosed. Readdy was perhaps counting on the fact that his audience would not be fully informed about the events that he was using in his argument. Or perhaps he hadn't thought through all the implications himself.

When the fire occurred, one *Soyuz* crew had been in the process of replacing another, and there were six men aboard, including one passenger, a German scientist. He did exactly what he was supposed to do: He put on an oxygen mask and stayed out of the way. Dennis Tito would have been able to do the same, and his presence would not have added to the hazard.

Readdy said that the collision had occurred "during a routine *Progress* docking." This was an inaccurate description of the event; it actually occurred during a very unusual test. But more to the point, the docking would never have been scheduled during a crew handover, when a passenger was on board, because during that period, there aren't any docking ports open. On the schedule of future *ISS* dockings, looking ahead five years or more, there isn't a single *Progress* docking of any kind scheduled during the time when two *Soyuz*es are docked to the station. Such a collision would therefore be impossible with a "nonprofessional" visitor aboard *ISS*.

Looking at NASA's claim in the light of all the details of the incidents, there is nothing to support the agency's argument. The accidents show that the presence of a visitor added nothing to the hazards, even in the worst-case situations.

On the issue of flying inadequately trained space travelers, Readdy was asked to defend NASA's history of flying its own passengers on shuttle missions—for example, the two congressmen in 1985–1986. The first had been Jake Garn, the Republican senator from Utah. Readdy assured the TV audience that Garn had the background to react properly to an emergency because "he was a former jet fighter pilot." When Readdy's statement was challenged after the show, he admitted that he'd misspoken; Garn had never been a fighter pilot, only a transport pilot. And Readdy hadn't even been asked about the other representative, Bill Nelson of Florida, who wasn't any kind of pilot.

NASA's public proclamations about safety being first stood in stark contrast to debates chronicled in internal NASA documents, which did not deal with Tito's flight at all. At a Safety Review Panel on February 28, two controversial subjects came up, fire safety and harmful noise levels. In both cases, there were serious doubts that NASA had learned or had properly applied lessons from previous space station experiences.

A safety official named Gregg Baumer raised a concern about the fact that program managers had removed a portable fire extinguisher from a March shuttle mission in order to make room for more cargo to be carried to the station. According to the meeting minutes, another engineer "questioned why the manifesting of hardware is given precedence above safety." When told that the decision had already been made and that it wasn't the business of the review panel, the engineer persisted. He "stated that the question is why the [extinguisher], which is defined as safety critical hardware, is at the bottom of the program's manifest list."

"Considerable discussion" ensued, noted the minutes. Experts asked "whether this sets an uncomfortable precedent." Another senior safety engineer "stressed that NASA as an organization has indicated safety has the highest priority, yet the management decision to demanifest the [extinguisher] implies safety is not the top priority." The issue was bumped back up to managers for reconsideration, but the outcome isn't documented. What's significant, however, is that it was even a subject for debate.

As for the hazards of excessive noise, the panel noted that Russia's short-term waiver for not meeting NASA safety standards had expired, and that nothing had been fixed. All the agency could do was extend the waiver.

An interesting pattern emerged from the first space station crew's debriefings and from the second crew's first detailed report. Some ground teams got kudos from the crews for operating smoothly, and some were criticized for poor procedures and inadequate support. One engineer noticed a correlation and sent out a memo on April 11, noting that the teams that operated smoothly were "on-board with this because of their experience during Phase I [shuttle-*Mir*]." As for the groups that were criticized, they did not have this experience and had evidently paid no attention to the volumes of "lessons learned" written by those who had.

However the new teams were learning the lessons they needed in order to operate a permanent space station, they were learning them, and it didn't take very long. Some shuttle-*Mir* veterans hit the ground running and operated well from the very beginning. Other groups struggled for a few months before evolving their own efficient techniques. Few had really learned much from shuttle-*Mir*, but the first months of the *ISS* showed that this really wasn't a big problem. The teams quickly learned what was needed, and they were able to adjust their unrealistic plans with ease.

We've encountered this theme before. The possibility remains that the previous U.S. experience on *Skylab* and shuttle-*Mir*, as well as all of Russia's experience with space stations, wasn't that critical after all. NASA's current workers proved themselves to be fast learners based on their own experience. Maybe we really didn't need the "Russian experience" after all.

But that opinion remains heretical. Not only is the emotional desirability of a partnership with the Russians frequently extolled, but such a partnership has been regarded as absolutely unavoidable. If you believe NASA officials, the Russian alliance didn't just help the *International Space Station* project, it was an essential factor in making it possible. In other words, it's widely claimed that NASA and its traditional allies would never have been able to create the *ISS* if the Russians hadn't been involved with critical aspects of the project.

I think that this is a slur on the American, European, Canadian, and Japanese engineers who had been working on space station designs since the mid-1980s. It is more than a little insulting to NASA's own space station workers, especially those who developed systems that predated the Russians' and that still form the backbone of the current *ISS* design. American space station technology leapfrogged the evolutionary but severely limited Russian space station assembly techniques, which advanced year by year in small increments. *Mir* was an elaboration of technological approaches introduced in the 1970s, and the two main Russian-built modules for the *ISS* are only minor improvements on that technology. In contrast, the equipment from the original *Freedom* team (the United States and its traditional international partners) is a generation or more ahead of the Russian designs.

The American hardware provides a great deal more electric power and communications capabilities than anything that the Russians were ever able to deploy. By using the shuttle itself as a workbench, the shuttle's robot arm—and later the even more capable station arm built by the Canadians—could move station modules and structures from place to place, attaching and reattaching the units as new ones arrived. These techniques allowed the modules to be equipped with normal-sized doors instead of tight hatchways (as on *Mir*), which in turn created the capability to move large objects (such as payload racks) from Earth to space and back again. Perhaps the boldest breakthrough on the American side of the *ISS* is the command and control system. This system is based on distributed processing by several dozen linked minicomputers, interfacing with the crew almost entirely through laptop computers and replacing heavy, inflexible mechanical switches and dials. All these technologies are based on designs by NASA and its original international team, created long before the Russians showed up.

Yet Dan Goldin never seemed to miss an opportunity to adulate Russian space expertise and deprecate American accomplishments. In 1997, Russian engineers threw together a special hatch door for the leaky *Spektr* module. It allowed the power cables inside *Spektr* to interface with cables leading to the rest of the complex. Goldin was so impressed that he exulted, "I tell you, my hat's off to our Russian colleagues. That's one incredible piece of hardware to put together in days. I wish we had the rapid reaction time in the United States that I see sitting in your hand there. We have a lot to learn from the Russians." NASA veterans winced at the slur, remembering—as Goldin may have forgotten—the crash development of equally impressive hardware to rescue the crippled *Skylab* space station in 1973. Time and again, Goldin repeated his litany about how the NASA team was made up of dunces who wouldn't know how to tie their own shoelaces on a space station if the Russians weren't along to share their secrets. Of course, when it came to the question about whether to fly Dennis Tito, Goldin decided that the Russians weren't quite so smart after all.

The Russians themselves take pride in their necessity to NASA. In discussing American intentions regarding the Russian alliance, *Mir* cosmonaut Aleksandr Poleshchuk told an August 1999 radio interviewer, "Their goal is our whole rich experience in manned space flight—this experience is very rich and the Americans are acquiring it at very little cost, they absorb it like a sponge." Added the cosmonaut, who is slated for a future *ISS* mission, "They could not create the Alpha station independently."

Yuriy Grigoriev, deputy chief designer at RSC-Energia, used the same metaphor to explain why the Russians were disregarding all American design suggestions (such as noise abatement) for the *Service Module*. In April 1999, he spoke to a Russian TV station. "There was no strong influence from the Americans because they have been actively absorbing our experience like a sponge over the last five years," he boasted. This was, he added, "because we have 30 years of uninterrupted experience and they just don't have that experience." (Space engineers often joke that there's a big difference between 30 years of experience and one year of experience repeated 30 times.)

Even with the hardware issues set aside, many still argue that the Russian presence was a necessary ruse to guarantee continued White House and congressional support in 1992–1993, when the *Freedom* project appeared to be stuck in a political mire. That's a fairly cynical view, in that it admits that NASA deceived the U.S. government with promises that failed in every case. The success of the project, in this view, was godfathered by fraud. If this is true, it hardly bodes well for future attempts to play the same trick for other big space projects.

In his dystopia *1984*, George Orwell defines the mental process called "doublethink" as the ability to hold two fundamentally conflicting opinions simultaneously. And there's no better example of space doublethink than the twin belief of many Americans that while the Russian presence did (admittedly) make *ISS* more expensive and slower to finish, it remains crucial to the major space projects of the future, which would be "too expensive" without the Russians along.

The "feel good" aspect of the partnership with Russia is equally frustrating. It implies that the only alternative to partnership is nuclear confrontation. NASA's official vision statement for the *ISS* reads: "A human outpost in space bringing nations together for the benefit of life on Earth—and beyond." The philosophical framework adds: "Our Mission—Safely build, operate, and utilize a continuously inhabited orbital research facility through an international partnership of governments, industries, and academia." Once upon a time, the research results of the project were primary, and the methods of achieving those results involved assembling an international team and keeping astronauts aboard the station continuously. But somewhere over the years, the means to the end became ends in themselves. So as Aldous Huxley once wrote, "The nature of the universe is such that ends can never justify the means. On the contrary, the means always determine the end." When international partnership became the only acceptable means to build the space station, in the end it became the primary rationale for the project.

As a result, if there is a philosophical theme to the *ISS* project today, it is that its success means the end of all major national space activities in the future. In this view, when it comes to human flight to the Moon or Mars, there should be no option for a purely U.S. project or for a U.S. project with traditional space allies. If the Russians aren't involved, the project should never occur. "What's really important is how we're doing it," exulted author Brian Burrough (*Dragonfly*) in a celebratory op ed on the occasion of the launch of *Expedition-1*. "This is humankind's station.....It's a real-life step toward a *Star Trek* universe, the first foray into The Federation." The science fiction metaphor was very apt.

Bob Cabana, the astronaut who commanded the first space station assembly mission in December 1998, voiced an almost theological passion for a permanent space partnership in a radio debate with me in mid-2000. "When we leave low Earth orbit, it's not any one country's responsibility, we need to do this united," he insisted. "If we can learn to work together 200 miles above Earth, in the vacuum of space, and pull this project off, we can do anything. And I think we're setting the stage for the future, and it would've been really wrong to do it without the Russians, without one of the major spacefaring nations of the world." With words such as "really wrong," Cabana let on that he had elevated philosophy above practicality.

At a prelaunch press conference at Baykonur, shuttle-*Mir* veteran astronaut Michael Foale asserted that "the model for space exploration is international cooperation." His strategy: "This flight is the keystone to all future exploration from this planet—to the Moon, to Mars and asteroids." Former astronaut Mike Baker, now a NASA official, agreed: "From now on, I think that all of our endeavors in space, human endeavors, will be joint."

Cosmonauts agreed. Yuriy Malenchenko stated: "This is how we in the Russian cosmonaut corps view the *International Space Station*: as a bridge to an international expedition to either the Moon or Mars." John Fabian, speaking before the House Science Committee in October 1993, voiced the same thought: "We are in a unique position to globalize human endeavor in space. . . . Cooperating with Russia gives the United States the opportunity to develop interdependent relationships."

Astronauts and cosmonauts are not the only ones to support the partnership (it is certainly possible that they are sincere, but they also are well aware that a lack of public enthusiasm for the partnership is a sure road to never getting another space flight assignment). Government bureaucrats said the same things. The U.S. State Department spokesman in Moscow, Nicholas Burns, told Interfax on August 7, 1998, "Our future in space is one of partnership with Russia. We have given up the space race, we have given up competition, and we're working together. . . . In the Cold War, we tried to compete with Russia. Now we try to put our efforts together, and that's a much better way of proceeding."

Speaking of his personal relationships with Russian cosmonauts, NASA space station commander-in-training Ken Bowersox told a press conference that "when you get close to [the effort], the emotions are so strong that Americans sometimes have to take a break from it." Bowersox then went on to the traditional false dilemma: "When you get that kind of relationship with people, you realize it is much better to be working together than building bombs and missiles." Dan Goldin made the same specious arguments: "Instead of pointing missiles at each other, instead of competing with each other, we learn from each other," he boasted, shortly after the launch.

"I've seen a change, not just in the Russians but in the Americans," he continued. "There was stress between our people, there wasn't trust." Then, thanks to joint activities over the past decade, things changed and mutual trust developed. "This trust is very important to do things. This trust is also a good sign for the future of the world." Goldin had trusted the Russians for years even as they continually misled and deceived him, and he wasn't going to give up now.

Once again, we're seeing the passionate belief that the example set by cooperating space workers will change the world. Dieter Andreson, senior space station manager for the European Space Agency, told a reporter, "There will never be strong conflicts between countries involved in the space station as long as we have astronauts for each others' countries on the outpost. That is one of my beliefs. And if it proves to be true, then it justifies not only the Russian delays in the program but the tremendous amount of investment the world is making in that bird."

Just about everyone agreed, it seemed. The project is "a test bed and training ground for large-scale international collaborations," wrote Dr. Robert Davies, a professor of physics at Seattle University who spent a year working in Moscow on technical liaison duties. "The vastness of the international collaboration [is] the station's preeminent value," he continued. "*ISS* becomes the metaphor for the challenges facing our planet in the coming century, and a model for tackling them. Global warming, mass extinctions, overpopulation, epidemics, pandemics—all are problems whose solution can be found only through consensus among all nations.... The skills acquired in the *ISS* training ground will not be lost to the far more serious challenges ahead."

These views reflect a backward interpretation of international diplomatic relations and joint space projects, as I have argued before. In the typical self-aggrandizement of space enthusiasts, experts and ordinary workers continue to repeat the mantra that friendship in space will promote friendly relations between nations. They continue to confuse the cause with the effect, potentially with dangerous consequences. As the *Apollo-Soyuz* Test Project (ASTP) showed, nations engage in joint space projects as a consequence of better Earthside relations, and as post-ASTP showed, they withdraw from such projects when diplomatic relations cool. The politicians and diplomats are not inspired by the space achievements; they exploit those achievements for shifting political purposes.

The participants in such projects see themselves as heroes forcing an amicable view of international relations upon unwilling political leaders. This is a very satisfying self-image, and potentially a very imprudent one.

Here's why. As agents of a foreign policy perhaps not shared by Washington, some space program workers could justify a wide array of actions, some of which are foolish or even illegal. The American impulse to perform acts of unselfish generosity in order to be liked has banged its head against the wall of reality for decades (especially with Russians), yet it still survives. In the aerospace sector, the technology involved makes this private impulse potentially dangerous. In an example I personally observed in 1995, when Russian engineers expressed an interest in acquiring U.S. equipment for laser ranging between space vehicles, the American side eagerly tried to find a way to get them the equipment.

They were frustrated and upset when they found that giving them this equipment would violate U.S. technology transfer regulations. Whether any of them went further in helping out their new Russian friends with this problem, I could never discover

At the extreme (and there is no evidence that space workers have ever been in this situation), we can now recognize the twisted idealism of certain Americans in the 1940s. While working on the Manhattan Project to build the atomic bomb, they took it upon themselves to lay the groundwork for a balanced world order by sharing technology secrets with the USSR. In hindsight, they described—and many still rationalize—their motivations as ethically superior to those of the U.S. government.

It doesn't take much imagination to see what is behind the impassioned and sincere speeches of space officials when they defend the Russian partnership as "changing the world for the better." At the very least, there is the potential that these officials will take unilateral measures to keep the Russians happy, with or without (or in violation of) the permission of higher government leaders. "Good intentions" pave star-crossed orbits in space, as well as unwise roads on Earth.

An explicit example of this occurred in early 2000, when NASA found itself in conflict with U.S. law with respect to payments to Russia. Congress had stipulated that no further cash be paid to Russia until the U.S. government certified that the Russian space industry had shut down aid to missile programs in Iran and other "rogue states." The White House was unable to obtain any such assurance, but NASA was confronted with a Russian space funding crisis that was bringing the *ISS* program to a halt. Arguably motivated by higher purposes, NASA decided to consider itself exempt from governmental constraints and chose to circumvent the law, using a provision that had been designed as a prudent exemption: This provision stated that if crew safety was involved, cash transfers to Russia were permitted. NASA simply declared that *all* human space-flight expenditures involve crew safety, and hence none of them could be limited by the law.

In another lamentable case, NASA medical officials had been dispensing about $20 million in American money to various Russian research laboratories to prevent them from collapsing. One group, called Biopreparat, was supposedly a pharmaceutical research institute.

But it turned out that many of its facilities and staff, as well as its director, General Yuriy Kalinin, had formerly been conducting Russian germ warfare experiments. According to many Russian experts, such activities continued to take place. The NASA official in charge claimed that he had never heard of Biopreparat.

"What happened in these cases was outrageous," a senior U.S. national security official told the *New York Times* in January 2000. "A.I.D. [Agency for International Development] and NASA were essentially running their own foreign policy." A former senior space medicine expert who had served in the NASA office that was involved in the payments told me that the officials in charge were "not trusted" and that "the whole process was tainted, but it would be impossible to prove." He concluded that "they really ought to get someone honest, competent, and credible as head" of that department at NASA.

From the start of the partnership, the Russians had warned NASA not to insult them by hiring "former spies" (defined by them as anyone too familiar with Russian culture or Russian space activities). But it soon appeared that there were plenty of strange characters who seemed out of place on the Russian side. Unlike the typical senior Russian bureaucrats, who were overweight, paunchy, and pale, these guys were fitness fanatics who pranced around energetically. Their English was also extraordinarily good. But it was difficult to keep track of them because NASA centers, I was told, often refused to support the FBI by giving it access to personnel and records. Even ex-astronauts in management positions outside NASA were ordered by their former NASA bosses not to cooperate with the FBI so as to avoid offending the Russians, according to two of them who talked to me.

At one private meeting, top Russian and American space officials were toasting each other in order to get better acquainted. One of these "out of place" Russians kept on filling everyone else's shot glasses, but as he tipped the bottle over his own glass, he slipped his thumb over the opening.

One of the Americans who could still see straight spotted the trick and laughingly protested. "Aha, Aleksey!" he shouted, "That's an old KGB trick!" The Russian didn't laugh or throw the joke back at his semiserious accuser; instead, he grimaced and blushed. His boss didn't laugh, either; he scowled at the Russian trickster darkly.

There were many reports of the same official being caught wandering around NASA sites on weekends, trying to see how far he could get without the escorts who were supposed to accompany him. But being pushy and nosy is a far cry from being proven a spy.

Leading U.S. opinion makers still seem to harbor delusions about other national security issues. Take the *New York Times*, the "gray lady" of the U.S. East Coast press establishment. The newspaper's editorial page had been "lukewarm about the station from the beginning," expecting it to yield "only limited scientific returns." In addition, NASA's promises "have been hyped beyond reason and debunked by expert committees," it wrote on the occasion of the first permanent crew launch. The newspaper was not wholly negative; it liked the symbolism of 16 nations collaborating. The only other value it perceived was "the participation of Russian space and rocket scientists who might otherwise be tempted to put their skills to malevolent use." The harsh reality of several hundred thousand unemployed missile engineers, and the ease with which any nation can hire them for work either in Russia or overseas, were blithely unrecognized by the *New York Times*.

There was one other angle of the Russian partnership that really gnawed at me, to the point that even my friends thought that I was being obsessive. If NASA officials thought that they were above normal security regulations, they also seemed to behave as though they were beyond moral considerations. In a project touted as bringing peace to Earth, one requirement was to avoid thinking about war—in particular, the war in Chechnya. The first war, in 1994–1996, was bad enough, but at that point NASA had little influence on Russia. In 1999–2001, while the second round of the terrible war raged in Chechnya, Russian soldiers engaged in acts against the civilian population that were so hideously cruel that they would be designated war crimes in any nation weak enough to be bullied (like Yugoslavia).

Curious about reports of immense clouds of smoke from dozens of oil well and refinery fires that had been set and left to burn for more than a year in Chechnya, I began to search for the "Earth view" photographs from shuttle missions that would show signs of the devastation. Hundreds of shots of surface targets all around the world, including Russia, are routinely taken from shuttle windows, both as part of a menu of desirable locations on daily schedules and as impromptu opportunistic shots. But I soon found that there were no shots of Chechnya, at least none that I could locate. I was given any number of technical excuses, but I suspected a larger one. For the sake of a project that was deemed eminently important to the fate of the world, space workers may have been willing to avert their eyes from such ugly realities, to pretend not to see what was arguably genocide against a colonized nation.

Motivated by the "higher purpose" of generic peace on Earth, many of the space workers I spoke with about Chechnya acted like moral eunuchs. They felt that the Russians' actions were too minor to worry about ("And don't forget what we did to the Indians" was a common response). I would argue that this alone was too high a price to pay for the debatable benefits of the partnership with Russia.

In any case, at this point (mid-2001) it seems as though the metaphorical "marriage in the heavens" between the United States and Russia could evolve along as many different routes as there are different kinds of marriages on Earth. Mutual interdependence could still lead to productive cooperation and fruitful results. An unequal flow of support could lead to resentment and unfulfilled dreams, even if the formal relationship is maintained. Or the alliance could end in divorce, either amicably, through a growing difference of goals, or bitterly, through betrayal.

The possibility of a breakup of the grand space alliance and a physical dissolution of the *International Space Station* may come as a shock to many observers whose enthusiasm was buoyed by the belated but undeniable success of the initial station assembly. But the *Zvezda Service Module*, the Docking Compartment, and even the *Zarya FGB* are not really contributions to the project, they are merely loans—permanent loans, it is hoped, or at least long-term loans, but always subject to foreclosure and repossession. The Russians will always have the option to recall them from the joint project and operate them on their own, in a separate orbit.

Even during the wide-scale Russian anguish over the termination of *Mir* in early 2001, when politicians, ex-cosmonauts, and space experts loudly lamented the loss of an independent Russian manned space program, nobody pointed out that there really *was* a successor to *Mir*. "*Mir-2*" was already in space, hooked up to the *International Space Station*. It remains the property of the Russian government, however, and it remains fully capable of independent flight.

Unhooking *Mir-2* would not be difficult. Space-walking cosmonauts would have to detach a number of external cables. Then the hooks and latches that keep the two sections locked flush together would have to be commanded to open. After that, the physical separation would be complete. The two spacecraft would then drift apart, and the renamed *Mir-2* could fire its thrusters to enter a new orbit at a different altitude from that of the rump *ISS*. Whether *ISS* could survive such an amputation is problematical.

The only plausible rationale for such a scenario would be diplomatic developments back on Earth. The *ISS*, after all, reflects Earthbound realities. The cosmonauts are employees of the Russian government, not of the international partnership. And if any Americans on the station are tempted to argue, they shouldn't forget that the Russians have the only gun.

Any prediction about the future of the Russian space alliance must depend on an understanding of the cold-blooded calculus of Russian motivations for remaining partners in the *International Space Station*. Like any rational nation, Russia will remain in the program for as long as it is in its national interest to do so, and no longer.

Currently, the balance sheet shows numerous advantages. "Within the framework of the program's implementation we gain access to all the results irrespective of where they are produced," Russian space program director Yuriy Koptev told *Rossiyskaya Gazata* in 1998. "Each partner has access to them by agreement. We are thereby substantially enhancing our technological level."

And it's a bargain. "We have the right of the permanent presence of three cosmonauts out of the total crew of seven," Koptev told a Moscow press conference that same year. "We have 35 percent of all resources on that station, while paying 6.8 billion [out of] overall cost in the range of 100 billion," referring to the dollar costs. "This is a tribute to our experience, a tribute to our luggage that we brought to the project," he added. Koptev was referring to the "Assembly Complete" configuration with a seven-person crew. That won't arrive for a number of years, but in the intermediate configuration (what wags are already calling "Half-Assembly Complete"), the permanent crew is three (on a normal four-month orbital duty tour). Russia is allocated on average half of those slots.

These numbers show that Russian participation in *ISS* is a good deal for the Russians. They can earn several hundred million dollars a year, more than enough to cover their expenses, by selling tickets. First, they can sell some of the seats on the *Soyuz* "swap" missions, currently two flights per year, at perhaps $15 to $20 million each. Also, they can sell some of their four or five annual allocated seats on long expeditions to other international partners seeking expanded research opportunities, for perhaps $80 to $90 million each.

Koptev also described another reason why Russia needs to stay in the *ISS* project. He pointed out that Russia's failure to meet its obligations on *ISS* would be "catastrophic." He explained: "Once we ruin the *ISS* project, we will never get an access to the international market. The resources that today support Russian space studies will vanish." So in Koptev's view, the *ISS* partnership is the "cost of doing business": It encourages Western governments not to interfere with the flow of more than half a billion dollars a year into the Russian space industry. Since it is the Russian space industry and not the government which mainly profits from such sales, government officials remain reluctant to spend federal tax money for a project that benefits these industrial entities, and have demanded that the industry itself fund the *ISS* support from the profits of these sales (probably half to two-thirds of the commercial cash flow is profit).

Based on these considerations, the Russian government has good reason to remain a "partner" in the *ISS*. In order to keep commercial programs running and Western cash flowing, the Russians can be expected to promise whatever it takes, while performing the absolute minimum.

A more optimistic view persists, however. It was recently voiced by Gene Kranz, a former flight director who left NASA when his prescient warnings about the Russian partnership were ignored.

"The space station's current problems and cost overruns do not reflect a failure of NASA technical management, but a failure of political leadership," wrote Kranz in February 2001. "NASA's problems with the space station for the better part of the last decade are the responsibility of Daniel Goldin and the questionable top-level leadership he selected during the re-baselining and initial design of the international space station (*ISS*).

"The costs faced by *ISS* program management in the year 2001 are the direct result of the technically and politically inept decisions in re-baselining the program in 1993–1994," Kranz continued. "Goldin embraced the Gore-Chernomyrdin initiatives and strove to establish Russia as a partner in the space station program, ignoring the technical and economic consequences of his act in a successful gambit to save his own job."

Kranz described the decisions that were made over his own objections, objections that led to his sudden departure from NASA: "Russia was subsequently assigned partnership responsibilities for critical in-line tasks with minimal concern for the political and technical difficulties as well as the cost and schedule risks," he explained. "This was the first time in the history of manned space flight that NASA assigned critical path, in-line tasks with little or no backup."

Nor had Kranz been alone in advising Goldin that his policy was foolhardy and delusional. Goldin knew that all of his experienced technical managers were against the policy, so, Kranz continued, "he bypassed them and established a redesign team headed by astronauts Bryan O'Connor and Bill Shepherd—neither of whom had relevant program management experience. As a result, the team he formed was inexperienced in program management, design requirements, systems and operations integration, and cost assessment."

Throughout 1993, Kranz and his associates wrote a series of memos warning of the likely consequences of such policies. Finally, it seems, Goldin had had enough. "On Sept. 17, Goldin dispatched his associate administrator for space flight, Gen. Jeremiah Pearson [an outsider Goldin had brought in to carry out his orders], to the Johnson Space Center, to personally deliver the message, 'no more memos.'" Goldin even brought in other military officers he knew from his industry days, whom he could count on not to get distracted by technical advice from the experienced space engineers.

The results were as Kranz and his colleagues had warned. "Today's problems with the space station are the product of a program driven by an overriding political objective and developed by an ad hoc committee, which bypassed NASA's proven management and engineering teams," Kranz concluded.

But by early 2001, Kranz was more optimistic. Most of Goldin's special lieutenants had moved on to more lucrative jobs in the aerospace industry, and Kranz believed that NASA veterans have been able to rescue the faltering project. "In the last two years," he recently asserted, "station management at Johnson Space Center has reverted to its more traditional technical and managerial roots and is making remarkable progress toward recovery. Today's program management team and the NASA field centers are on track to resolve the budget issues and complete the station program." That was before the discovery of another $4 to $5 billion dollars in cost overruns.

Regardless of one's viewpoint, the United States and Russia have major roles to play in space—in national programs, in alliances with junior partners, and in partnership with each other. There will be great space projects beyond *ISS*, involving the construction of bigger space facilities. Human expeditions beyond low Earth orbit will be launched. Such projects will be conducted by partnerships of nations or by individual nations, as determined by the national interests of each government.

The extent of Russia's future role will be determined both by the lessons learned from the *International Space Station* and by the persistent power of the illusions and wishes supporting its role. The success of the projects will be determined by the balance between these two opposing themes.

In March 2001, a discussion about Russia's failure to become a democratic, lawful, and peaceful nation took place at the United States Institute of Peace. Dr. Anatol Lieven, a senior associate of the Carnegie Endowment, made an observation that applies both to Russia in general and to Russia's role as a partner in the rest of the world's space activities. "The problem is that we set hopelessly unrealistic expectations for Russia," he stated. "This is a very old and dangerous tendency in the West. Then when they fail, we don't ask what is possible in practical terms, but we insist upon thinking that they're innately wicked."

There are plenty of optimistic scenarios in which international programs are strengthened and grow more valuable. Earth won't be adequately served by a single space station, and in the coming decades, others may appear that are purely Russian (perhaps in polar orbit), purely Chinese, purely Microsoft or Hilton, purely Mormon or Lamaist, or serving manifold combinations of players. Once we figure out how to do it, expeditions to the Moon and the planets may also fly under different flags, including flags not yet drawn. Russia has a role to play in human expansion into space, and no single strategy for its activities has ever been permanent.

Speaking as a ballistician (an orbital mechanics designer), I see the spacefaring nations of the world as traveling through the years in measurable orbits. An orbit around Earth is a closed path through space that crosses different regions as the Earth turns, but always winds up back at the same point, ready to retrace its route again and again. Many lamentable examples illustrate a comparable repetition in our relationship with Russia, driven as we are by mutual misunderstandings and illusions. The stars surround us, but our pathways have been truly star-crossed. Despite our apparent motion, we have made no measurable progress relative to the stars.

In the world of real space flight, to break free from a repeating orbit requires an "escape velocity," an application of extra energy and careful guidance that leaves the old pathways behind, opening up the Universe. That's the metaphor I think we need, the star we need to steer by. If we ever hope to venture beyond Earth orbit, we must first break free of the star-crossed orbits of misperception that bind us to the ground. Sent by individual nations or alliances of nations or partnerships unimaginable today, these future human expeditions will require energy and guidance that we have so far been unable to muster.

Notes

Introduction

The CNN *Crossfire* program with Graham and Torricelli was broadcast on September 7, 1988; supplemental information was given verbally by the *Crossfire* producer. The early politics of the Space Age are described in Walter McDougall, *The Heavens and the Earth*, Basic Books, New York, 1985, and in my own *Red Star in Orbit*, Random House, New York, 1981.

Quotations on the Baykonur geographic deception are from John Rhea, ed., *Roads to Space*, McGraw-Hill, New York, 1995. The Russian space records folders were being offered for sale by Kaller's America Gallery in New York City.

Russian accusations that the space shuttle was used for weapons testing were originally compiled in my book *The New Race for Space*, Stackpole Publishers, Harrisburg, Pa., 1984, pp. 145–152, from contemporary Soviet press accounts translated by the Foreign Broadcast Information Service (FBIS), Washington, D.C.

Dr. Paul Spudis's paper on the impact on Soviet strategic thinking of the *Apollo* success was shown to me in a prepublication draft, with permission to quote. Jeffrey Manber's insightful study is "Russian-American Space Miscommunication: A Study in Missed Opportunities," *Space Policy*, January 2000, pp. 3–6.

Chapter 1: Zenith

Accounts of the *Salyut-7* rescue come from contemporary press accounts and from relayed personal communications from Feoktistov. *Aviation Week*'s prediction of Soviet space advances was in the March 12, 1984, issue.

The Karen Blixen quotation was not actually written by Isak Dinesen, but was written by Kurt Luedtke for the film *Out of Africa*, directed by Sydney Pollack, Universal Studios, 1985.

Dzhanibekov's character was described to me by ASTP personnel who knew him. We have never met. The first Western major media news treatment of the rescue was by Warren Leary in the *New York Times*, June 24. The first Soviet account "Courage of the 'Pamirs,'" by Konstantin Feoktistov, was in *Pravda*, August 5, 1985. Additional biographical information on Feoktistov came from personal communications from the Belgian space historian Bart Hendrickx and others.

Chapter 2: History

My *N-1* rocket fragment is a prized possession, and I am eternally grateful to those who provided it and those who transported it, all of whom prefer to remain unnamed. The visit of Smetanin's team to Houston was organized by space lawyer Arthur Dula, who fits the classic definition of a true pioneer: the guy you find face down on the trail in front of you with an arrow in his back. Other corporate entities took over the international commercial rocket deals he spearheaded. Dula also hosted the Moscow restaurant meeting where I was first told the "secret name" of the *N-1* rocket. Semyonov's visit to the Johnson Space Center was for a speech at a space conference on October 21, 1991.

Most of the Russian retrospective quotations are from Asif Siddiqi's monumental chronology of the Russian side of the Moon race, *Challenge to Apollo: The Soviet Union and the Space Race, 1945–1974*, NASA History Division (SP-2000-4408), Washington, D.C., 2000. This is—and probably will forever remain—the single authoritative English-language account of that project. There are shorter overviews, with some different details, on the best-in-the-world space-flight Internet site, Mark Wade's *Encyclopedia Astronautica* (http://www.friends-partners.org/~mwade/spaceflt.htm).

Information on the machine gun installed on the *Salyut-3* space station in 1974 and on the 1987 space battle station called *Polyus* was compiled from a dozen sources during e-seminars on the space discussion group hosted by "Friends and Partners in Space."

Chapter 3: Decaying Orbit

The late-1980s trends toward major space spending cuts were chronicled in my 1989 article, 'Trouble in Star City,'" *Air & Space Magazine*, from which many of the Soviet quotations (those from Avduyevskiy, Kerimov, Shatalov, and Leskov) are taken. For example, Shatalov's interview was in *Nedelya*, Oct. 12–18, 1987.

Mass layoffs in the space industry were described throughout the Russian media. The 1997 "open letter" was published in *Sovetskaya Rossiya*, Oct. 18, 1997, p. 5. Dr. Twygg's comments come from her prepared remarks for the September 1998 congressional hearings on Russian space infrastructure, at which we both testified.

The Baykonur workers' petition was reported by the ITAR-TASS news agency in Moscow on its English wire on Dec. 17, 1993, from correspondent Vladimir Akimov. The court sentences on rioting soldiers were reported in a Dec. 13, 1993, Associated Press story. Belgian space expert Theo Pirard wrote about the population crash for *Satellite News* (Sept. 5, 1994). Theft of wire was described in the Almaty, Kazakhstan, newspaper *Ekspress-K* on Oct. 6, 1994, by reporter Georgiy Loriya. Problems of local law and order were described by veteran Moscow space journalist Sergey Leskov in his story, "Baykonur Recruits Its Own Army outside the Defense Budget," *Izvestiya*, Dec. 28, 1994.

Gore's statement on Baykonur's health, and the ANSER corporation's report on which it was based, are described in contemporary press accounts. Also, I was able to obtain a copy of the report's executive summary. The camp-fires were described in *Rossiyskaya Gazeta*, June 21, 1996.

Bulgakov's dictum, "Write what you see, and do not write what you do not see!" was quoted in a newspaper article written by Baykonur Cosmodrome chief Aleksandr Shumilin, entitled "Rockets Still Blast Off—What Lies in Store for Baykonur?" published in *Rossiyskaya Gazeta*, Moscow, on Feb. 1, 1994.

Chapter 4: International Orbits

NASA's official history of the first U.S.-Russian space docking is *The Partnership: A History of the Apollo-Soyuz Test Project*, by Edward Clinton Ezell and Linda Neuman Ezell, The NASA History Series (NASA SP-4209), Washington, D.C., 1978. It is an outstanding piece of work. It also won a "Golden Fleece Award" from Senator William Proxmire for costing far more than was expected, but it was still a bargain.

John Logsdon's paper on the fallacy of functionalism is "U.S. Soviet Space Relationships in the 1990s: A U.S. Perspective on Policy Alternatives," IAA-88-624, given at the 39th conference of the International Astronautical Federation, held in Bangalore, India, in October 1988.

The collapse of post-ASTP cooperation and the tentative and stillborn moves to restart joint activities were chronicled in "A Shuttle-*Salyut* Joint Mission," chap. 12 (pp. 132–144) of my book, *The New Race for Space*, Stackpole Publishers, Harrisburg, Pa., 1984. The case for renewed joint manned activity was described inter alia in my articles "US-Soviet Space Missions Can Start Soon" (*Christian Science Monitor*, Oct. 2, 1989, p. 19) and "Russian Lessons: Can the Soviets Teach NASA How to Build a Better Space Station?" (*OMNI*, December 1990, p. 26). Another *OMNI* article of mine calling for a shuttle-*Salyut* docking was reprinted as 'Hitch Up with a Red Star," in Frank Kendig, ed., *The OMNI Book of Space*, Kensington Publishing, New York, 1990, pp. 251—259.

The discussion of docking mechanisms is adapted from my article "United We Orbit" in *Air & Space Magazine*, February—March 1996.

International missions to *Mir* are chronicled in Rex Hall, ed., *The History of Mir 1986—2000*, British Interplanetary Society, London, 2000.

Post-ASTP discussions for renewed U.S.-Soviet space cooperation were discussed in a chapter of my 1984 book, *The New Race for Space*," from which the Beggs quotations are taken. Dzhanibekov's put-down of the accuracy of the space shuttle's flying was in an interview in the *Denver Post*, Apr. 7, 1987, during a visit to the United States. My own paper on near-term incremental space cooperation options was delivered at the "Case for Mars IV" conference in Boulder, Colorado, on June 5, 1990, and published in the proceedings, Tom Meyer, ed., *Case for Mars IV*, UNIVELT, Inc., San Diego, 1997.

Leonov's recollections of ASTP, and his warped recollection of the U.S.-Soviet rivalry, were expressed in Yuri Salnikov, "ISS Mission Carries the Spirit of *Apollo-Soyuz*," *Moscow Times*, Nov. 1, 2000.

Chapter 5: Origins of the Partnership

The narrative on the genesis and early life of the Russian partnership is adapted from my article "NASA's Russian Payload" in *American Spectator*, August 1998, pp. 38—45, from which the quotations were drawn. Jeff Manber's comments are from his article "Russian-American Space Miscommunication: A Study in Missed Opportunities," *Space Policy*, January 2000. The early considerations were well summarized in "Space Station: Update on the Impact of the Expanded Russian Role," GAO report B-257996, July 29, 1994, by Donna Heivilin, Director, Defense Management and NASA Issues, to Senator William S. Cohen, Committee on Governmental Affairs, and in "Report on the Task Force on the Shuttle-*Mir* Rendezvous-Docking Missions," July 29, 1994, by a task force of the NASA Advisory Council, headed by Thomas Stafford. The hyped case for the partnership is exemplified in "International Space Station," a 50-page May 1994 NASA booklet.

Quotations from Goldin, Gore, and other government figures are from contemporary press accounts (e.g., Goldin on "closed the books on the cold war" was in the *New York Times*, Sept. 3, 1993). Comments from Bryan O'Connor and from other NASA officials, who requested anonymity, come from personal interviews. Insights were also shared by Nick Fuhrman, a former House Space Subcommittee aide, in a long interview on May 19, 1998. Quoted comments from NASA internal briefings are from my own contemporary notes of the meetings.

Chapter 6: *Mir* Breakdowns

Linenger's first-hand accounts of the fire are taken from private e-mail he sent down to NASA, which I have copies of. He also provided insightful descriptions for a BBC-2 program (*"Mir* Mortals") which aired the following year, on April 23, 1998.

I had access to all the on-board photography at NASA, but several of the images (which showed ointment-dabbled hands and chests) were refused public release on the grounds of medical privacy, allowing NASA officials to issue inaccurate statements on that subject ("nobody was injured") with no fear of contradiction from the evidence they withheld.

Goldin's and Culbertson's public comments are from contemporary press reports. The comment on how "one *Soyuz* was blocked, but the other one was not, so the lives of the astronauts were not in danger at any point" was from CNN, June 27, 1997. Goldin's assurance that the fire was "easily manageable" was in a statement to Congress on June 18.

Stafford's denial of any previous safety problems was from his written testimony submitted to Congress. Another similar claim, "2500 Candles Burned without Incident," was in a memo from Stephen Tripodi, JSC-DF82 (EECOM), March 19, 1997, re "*Mir* Status Meeting."

The official Russian report that denied any previous fires was "An Analysis of Off-Nominal Situations Associated with the Service Module (SM) of International Space Station Alpha," NPO Energia, September 1994, SS346/TTI/10/3/94/AM, p. 80.

Other reports of the *Salyut-1* fire were collected in interviews by Bert Vis and other European researchers. Chretien's account was given in a 1988 NASA meeting which I attended and documented in my own notes at the time. The Oct. 15, 1994, fire was documented in *Novosti Kosmonavtiki*, and my article on the issue appeared in *Space News*, Apr. 12, 1995. NASA's inadequate knowledge of these events was the subject of a long private conversation I had with a NASA safety manager on March 21, 1995, which I wrote down within an hour. The accounts of Russian comments to the ABC News team came directly from the segment producer, in personal communications with me. Cushing's Inspector General memo dated Oct. 15, 1997, was posted on the NASA-OIG [Office of the Inspector General] Web site.

Culbertson's denial of any knowledge of previous fires was in an ABC 20/20 news story, which aired in April 1998. The questionable claims about fire safety lessons from the incident were researched as part of an article for *Spectrum* magazine. Quotations from internal NASA documents are based on copies of the actual documents in my possession. Wolf's comment on no fire drills was from a report, 'Safety Debrief,' posted on *NASA WATCH*, dated Apr. 25, 1998.

Chapter 7: *Mir* Screw-Ups

The recollections of the collision with *Mir* of Culbertson, Lazutkin, Foale, and Tsibliyev are collected from contemporary press accounts and from interviews for later television documentaries such as those by the BBC and PBS *Nova*. There were also details in Foale's private postflight debriefing with NASA that weren't ever released.

Comments from Bogdashevskiy, Urinson, and others were taken from contemporary press accounts, translated by the US FBIS. Culbertson and Ryumin exchanged a series of frank official (internal) messages, of which I have obtained copies. Culbertson and van Laak were quite candid during an interview I conducted with them on Jan. 16, 1998, for an article in *Spectrum* magazine.

I received other comments and other internal documents from a wide variety of sources within the space communities in Houston, Washington, D.C., and elsewhere.

The July 3, 1997, memo was an example of a report that was widely circulated within NASA for everyone's information, as per NASA's old standards of maximum information exposure, but was then ordered restricted under the new policy of minimum disclosure of minimum information even within the space team.

My own report on safety issues, filed on Oct. 12, 1994, argued against Russia's assurances that docking was a safe operation that could never result in depressurization of the station. Based on safety practices I had been taught by NASA old-timers, I argued that the safety of such operations remained unproven.

News about multiple collisions (not the single collision which NASA reported) of the *Progress* with the *Mir* in June 1997 reached me beginning on Aug. 4 from several colleagues still working inside NASA. When asked at a press conference, Culbertson confirmed these reports, but had shown no inclination to volunteer the information. Navias's comments were made at the same press conference.

Details of the *Inspektor* project and its hazardous failure in December 1997 are from contemporary press accounts, from the project's Web site, and from internal NASA reports which fell into my hands. The December 2000 comment from the first *ISS* crew that they needed more windows in the station was overheard by me on the air-to-ground link (NASA subsequently shut off public access to the crew's conversations "as an economy move").

Young's comment that the accidents were not age-related was made at a Sept. 25, 1997, press conference. A copy of the minutes of the Stafford Committee meeting in Denver on July 23–24, 1997, shows they knew this wasn't true.

The July 17 "Lazutkin-pulls-the-cable" loss-of-control incident is described in contemporary press accounts. Engelauf's memo, which tells the real story that both NASA and the Russians wanted suppressed, was dated July

18, 1997. His comments to me about the severity of his criticism were made at a press conference on Jan. 13, 1998.

My December 1995 article in *Spectrum* magazine was called "Russia's Space Program: Running on Empty." Culbertson's comments to the Stafford Committee are from minutes of the meetings that were provided to me by a friend. Tsibliyev's explanation of the flood of foul-ups was given at his postlanding press conference, reported in contemporary Russian and Western press accounts. Blagov's comments were made in interviews with a major broadcast network newsmagazine in December 1997 for a segment that never aired, but for which I was shown the transcript.

Blagov and Solovyov's statements of cocky confidence that they could keep *Mir* going forever were quoted in late-1997 Russian press accounts.

Chapter 8: The *Mir* Safety Debate

The landing of the *Soyuz* in August 1997 comes from contemporary press accounts, including the postflight press conference. Prepared statements, contemporary notes and press coverage, and a transcript of the Sept. 18, 1997, congressional hearing on *Mir* safety are also available for the comments of Marcia Smith, Roberta Gross, Frank Culbertson, Ralph Hall, and myself.

Jim van Laak e-mailed me some additional technical details about the cause of the crash landing on Apr. 27, 1998, which minimized any danger from such malfunctions. In contrast, the term *catastrophic* comes from a Russian safety document on *Soyuz* systems which I reviewed for NASA in 1997 (and kept a copy of).

Comments from Foale, Gibbons, and Lawrence that express gratitude for the problems on *Mir* are from contemporary press accounts. They are balanced by the long privately circulated essay on space flight safety by Charlie Harlan, the retired head of the safety office at the Johnson Space Center. Blagov's comments were made during interviews with an American news team for a segment that was never aired.

Hammond's comments were made in several internal NASA memos that he wrote in July 1997 and in an interview that he gave to the AP that was printed on May 28, 1998. Cabana's e-mailed rebuke to Hammond was sent on July 1, 1997. The story of the stolen voice tapes came from an individual involved in the investigation.

James van Laak's comments in the *New Yorker* appeared on the last page of Peter Maas's article "Messages from *Mir*," *The New Yorker*, Oct. 20–27, 1997, pp. 238–247. It's a well-researched article, but it pushes the hype that without shuttle-*Mir* experience, it would have been hopeless for the United States to attempt any new human space project. The briefing in which Rutledge stated that "there is no hard evidence that *Mir* is currently unsafe" occurred on Sept. 10, 1997. Wilhide's similar assertion was in an AP piece on May 28, 1998.

Utkin's failure to discover the cause of the loss of *Mars-96* in November 1996 is documented in my article "The Probe that Fell to Earth," *New Scientist*, Mar. 6, 1999. Merrill's opinion was expressed in letters he wrote to the Stafford Commission, copies of which fell into my hands, as did Linenger's actual reports from aboard *Mir*. Newkirk's critique of Stafford's conclusion was in a special report posted on his Web site on September 26, 1997.

The survey on *NASA Watch* about space workers' fears about disagreeing with management was conducted in late 1997 and can be accessed in its entirety at Keith Cowings' web site (www.nasawatch.com). Goldin's comments on the issue were from a June 1, 1998, senior staff meeting at NASA Headquarters. Culbertson's quote is from the *Washington Post*, Sept. 16, 1997. Ryumin's comments are from *Aviation Week* magazine, Sept. 22, 1997, p. 34.

Banke's comment on how critics were afraid was in his article in *Ad Astra* magazine, September–October 1997, "Moscow, We Have a Problem," p. 35. NASA's internal document, "*Mir* Safety Assessment Briefing" from Culbertson, was from August 1997, and Ryumin's warning against abandoning shuttle-*Mir* is from a Sept. 12, 1997 internal NASA memo, both of which fell into my hands. My public concern over *Mir* safety was the basis for my article, "NASA's 'Can-Do' Style Is Clouding Its Vision of *Mir*," *Washington Post*, Sept. 28, 1997, Outlook section.

Chapter 9: Rescue and Recovery

Foale's comments about Solovyov came to me from somebody who heard them, not from Foale. Thagard recently confirmed to me personally the report I'd heard about his declining to take part in the prelaunch ceremony involving the bus tire. David Wolf's comments are quoted from an interview I did with him in 1999 for *American Legion* magazine, and I also quote from a few of his e-mails from orbit.

The issue of the threat of hull degradation and loss of pressurization is based on numerous interviews, NASA internal documents, and consultation with an auto mechanic familiar with the effects of ethylene glycol on aluminum block automobile engines. The incident of rocking the *Mir* modules back and forth was told to me by an eyewitness who requested anonymity. The October 1983 hull patching story is from a 1997 book, in Russian, providing a historical overview of the Khrunichev Institute, which was given to me by a NASA engineer who visited there. James van Laak graciously provided me with a summary of the results of inspections of *Mir* hull corrosion in 1997–1998, but I couldn't obtain the raw reports, and my requests to interview the astronauts who conducted the inspections were refused. Ryumin's comments were from a press conference I attended and at which I made real-time notes.

Solovyov's extensive postlanding comments come from TASS, Feb. 21, 1998, and other contemporary Russian press reports. I somehow obtained a copy of NASA's internal debriefing reports for Wolf's mission (not from Wolf), and some quotes from him come from a profile that appeared in *Space News*. Dan Roam's story of Russian repair finesse was from a 1998 e-mail that was confirmed and cleared for publication in May 2001.

Information on Andy Thomas's experiences on *Mir* in 1998 come from personal conversations with him, plus contemporary press reports and a few NASA internal documents. The information on Salizhan Sharipov comes from several long interviews I conducted with him for *OMNI* magazine in January 1998. Information about other Russians who made space shuttle missions as guests of NASA came from insider friends who cannot be named.

Chapter 10: Lessons Learned and Unlearned

John Uri's memo on reading "*Skylab* lessons learned" was distributed in July 1999, and I received a copy from a secondary source. Comments from shuttle-*Mir* veterans such as Mike Foale and Dave Wolf were widely reported in contemporary news reports in 1997–1998. Wolf and Culbertson also talked directly with me for articles I wrote for *Spectrum* and *American Legion* magazines. Mike Foale's Internet answers were posted on Oct. 8, 1997. The Valeriy Ryumin quote is from an internal NASA report of a September 1997 briefing. I directly interviewed Max Faget and Walt Cunningham.

The description of NASA's efforts to capture "lessons learned" from shuttle-*Mir* is based on research for an article in *Spectrum* magazine (June 1998, "Shuttle-*Mir* Lessons for the ISS"). Interviews with Chilton, Lengyel, and Cardenas were on the record, while numerous other insiders talked with me for background only. Lucid and Tognini made their suggestions in interviews with journalists. I also obtained a wide variety of internal NASA reports and memoranda.

Goldin's and Trafton's comments were from contemporary press reports. In March 1998, NASA published its apologia, "The Phase 1 Program—The United States Prepares for the International Space Station." *Apollo* astronaut Jack Schmitt wrote his "op ed" on the subject for *Space News*, Dec. 1, 1997.

Chapter 11: *Soyuz* Secrets

The former astronaut's description of the early NASA studies of *Soyuz* as a rescue vehicle was in a long e-mail to me, May 10, 1999, followed up by telephone interviews. Although at first he was willing for me to use his name, as time went by he became more and more worried about retribution and in the end asked me to keep him anonymous.

General descriptions of the *Soyuz* spacecraft and its flight history can be found in Dennis Newkirk's book *Almanac of Soviet Manned Space Flight*, Gulf Publishing Company, Houston, 1990, and at Mark Wade's Internet site, *Encyclopedia Astronautica*.

The 1969 *Soyuz-5* near-disaster was described in the Russian news media (e.g., Mikhail Rebrov, "Difficult Return from Orbit," *Krasnaya Zvezda*, Apr. 27, 1996), and it also was mentioned on p. 184 of the massive 690-page official history, *Rocket-Space Corporation "Energia" Named After S. P. Korolyov (1946–1996)*, chief editor Yuri Semyonov, Moscow, 1996.

My private "Soyuz Landing Historical Reliability Study" was delivered to NASA on Mar. 19, 1997. The extent of NASA's official reaction was described to me in an e-mail from the contract monitor on Apr. 20, 1998.

Information on the *Soyuz* survival gun came from Americans who trained with it. I received detailed responses from Jim Voss, Dave Wolf, and Mike Foale (e.g., personal e-mail, Sept. 28, 1999).

Chapter 12: *ISS* in the Wilderness

The list of lamentations by NASA spokespersons in 1997–1998 came from many sources. Brinkley's comments were from internal NASA meetings and public press conferences that I attended and took notes on, as well as the May 13, 1998, issue of *Florida Today*. For example, his "doom and gloom" and "we feel a lot more confident" quotes were from an "All Hands" meeting at the Johnson Space Center on May 29, 1997. Lynn Cline's comment to ABC News was on July 16, 2000. Trafton's July 1996 comments were in congressional testimony, as was Jacob Lew's gobbledygook on Aug. 26, 1998. My friends within the project e-mailed me descriptions of the growing desperation and despondency and the obsession with secrecy and "no paper trail" decisions.

The collapse of the Russian *Altair* relay satellite system was chronicled in the popular press, but detailed accounts are based on research by Dutch amateur Russian radio eavesdropper Chris van den Berg, a retired engineer, provided to me in personal e-mail correspondence.

Goldin's 1998 quotes to Congress were from hearings in May, quoted in the May 7, 1998, edition of *Florida Today*. Rothenberg's quote about "a year ago" was from the *Washington Post*, Sept. 21, 1998. The comments from NASA's Englund in Moscow ("eventuate") were in a Nov. 16, 1998 story in Reuters. Sensenbrenner's comments were from the hearings in September 1998, as were Talbott's written responses dated Oct. 7, 1998. Clinton's late 1998 comment ("very important that we have … the Russians in the space station venture") was from *Florida Today*, Oct. 30, 1998. Boeing's Doug Stone is quoted from the May 1998 issue of *Popular Science*.

John Logsdon is quoted from *Space News*, Oct. 12, 1998, p. 21, and from a profile of him in the same newspaper (*Space News*, Nov. 17, 1997, p. 22). John Pike's comment on repair work is from a CBS/AP wire story, Sept. 22, 1998. Pike also appeared with me on the Diana Rehm talk show (NPR) on June 7, 2000, from which he is quoted.

Like so many others, my astronaut source at headquarters requested anonymity for fear of retribution. The comment on the challenges of an operational program "immersed in politics" is from a private communication on Apr. 29, 1998. Other comments are from face-to-face private conversations over the years with trusted insiders (who also trusted me) whose standing within the program would be threatened if their candor and honesty were revealed.

The GAO's report to Congress on Russian hardware hazards was entitled "Space Station: Russian Compliance with Safety Requirements." It was written by Allen Li, the associate director for the GAO's National Security and International Affairs Division. Former astronaut Blaine Hammond's comments were quoted from contemporary press accounts. Another astronaut's comments on serious noise issues with different equipment was from a private e-mail.

For the "mission creep" trap, Culbertson's quote on "a difficult time" is from an April 11 interview with *Space News*, published on Apr. 21, 1997, p. 18. Marcia Smith's softly worded demurral was in an AP story on Feb. 20, 1998. NASA's purchase of 4000 hours of Russian crew time was described in *Aerospace Daily*, Feb. 8, 2000.

The completion of the Indian rocket deal was described in Moscow Interfax on Sept. 21, 1998, and on the Delhi Doordarshan TV network on Sept. 24, 1998. The first launching of an Indian rocket with the cryogenic stage in April 2001 was widely covered in the Indian news media.

Chapter 13: Culture Gap

The letter from General John Deane in Moscow was published in Leonard Mosley, *Marshall: Hero for Our Times*, Hearst Books, New York, 1982. Michael Evans's book is called *TOPAZ International Program: Lessons Learned in Technology Cooperation with Russia*, Booz-Allen & Hamilton, Arlington, Va., 1994, under BMDO contract # SD1084-93-C-0010, 116 pages. His comments on the reaction to his "TOPAZ" report were from contemporary conversations, verified and expanded in a private e-mail message of Feb. 27, 2001.

Astronaut Bonnie Dunbar's comments were made in public in October 1994, at a regular monthly meeting of the Houston chapter of the American Institute of Aeronautics and Astronautics, and her anger about NASA's accepting blame for Russian delays is from an internal NASA e-mail from Dunbar to Kyle Herring on Sept. 22, 1994 (she is not the source). Judith Lapierre's experiences were described in the Canadian press (although not in the U.S. press) in March 2000, and my information was obtained directly from her by telephone for long articles for UPI and for GalaxyOnline.com. Follow-up conversations and e-mail provided additional insights. Relevant extracts from the contemporary Russian news media are thanks to FBIS.

Confirmation of Lapierre's assessment of poor Russian space medical data is from a private e-mail to me from a flown American physician-astronaut, known to me for many years, who for professional safety declined permission to use his or her name.

The dangers to Americans living in Moscow are detailed in a series of internal NASA memos which I have copies of. In response to inquiries from me about the worst beating case, I was also called by Mike Baker, a former astronaut who was head of NASA's Moscow office, and he provided a partial description of the event. For another example, the Dec. 26, 2000, warning memo to NASA employees in Moscow was from NMLO (NASA Moscow Liaison Office) Administrative Officer Patrick Buzzard at the U.S. Embassy in Moscow. Todd Breed's tragic story goes far deeper than described here based on published sources in the contemporary press (for example, see Todd Halvorson, 'Political Considerations Weigh Heavily in Russian Space Station Role," *Florida Today*, Dec. 22, 1996) and information passed to me by his close friends within NASA.

Other general cultural comments came from a variety of sources. Jorg Feustel-Buechl's comments on Russian technical culture were in an interview for Peter N. Spotts, "Europe Delivers on Space Projects," *Christian Science Monitor*, Dec. 21, 2000. NASA's own first encounters with the "no paper trail" Russian approach were detailed in internal e-mails from 1994–1995, which I kept copies of. Bruce Cornett's detailed description of Russian space operations culture is from a long e-mail on June 23, 1999. Foale's descriptions were for the BBC program "*Mir* Mortals" in 1999.

Less favorable aspects of Russian culture are documented both by my own personal experiences and in long philosophical messages from friends within NASA who have spent years in Moscow. The "lie, backstab, cheat" private e-mail was sent on Sept. 11, 1997, from a friend in Russia. Other details came from Steven D. Jones, the director of the East-West Business Strategies group in San Francisco, at www.ewbs.com. His quoted comments are from a personal e-mail, Jan. 19, 2001 (sdjones@ewbs.com).

Chapter 14: Staking Out the Orbit

The design of the *ISS* orbit was documented in a series of internal NASA memos in 1996–1997 and was the subject of my "sustained superior performance" award, given in May 1996. Russia's brief attempt to go back on these design agreements two weeks before launch was brought to my attention by an e-mail from a friend on Nov. 4, 1998, and was described in the news media and in follow-up e-mails over the next few days.

My attempts to locate any record of NASA's acquiescence to the Russian name *Zarya* for the *FGB*, even though it officially was an American module, were fruitless. Even requests under the Freedom of Information Act brought the response that there were no written accounts of the discussion and the rationale. My comparison of the premature launching of it to "the longest 'Hail Mary' pass in history" was during my congressional testimony on Oct. 7, 1998.

The secret near-disaster on launch day was revealed by the Russians in several Moscow TV programs on Aug. 4, 2000, almost two years after launch. There is still no reference to these events anywhere on NASA official Web sites.

Chapter 15: Follow the Money

My cash transactions with Klimuk and Glazkov were on behalf of Sotheby's auction programs. My much more unpleasant encounter with Vladimir Solovyov was in February 1995. My *Washington Times* article about the "cosmonaut mansions" had appeared on Nov. 24, 1994. The details of Byron Harris's ambush of Glazkov are from private interviews, and not all the details are ready for disclosure even now. His *Nightline* appearance was in April 1995. The dispute over the houses, which continued into my own congressional testimony in October 1998, was followed by Eddie Bernice Johnson's letter to me dated Oct. 26, 1998, which repeated White House assurances that the houses did not exist.

Other incidents of space-related corruption in Russia are from contemporary Russian press accounts and from personal communications from friends of mine within NASA. The financial suffering of ordinary Russian space workers is well documented in the Russian press from that time. For example, the space veterans' letter of Oct. 4, 1997, is quoted in *Sovetskaya Rossiya*, Oct. 18, p. 5. Numerous FBIS reports addressed this explicitly, including 200,000 rocket scientists' losing their jobs (*Trud*, Dec 10, 1994, p. 2), Yuriy Koptev's 1995 prediction of more layoffs in Moscow (Interfax, Jan. 26, 1995); the pitiful status of the Kuybyshev rocket factory (Moscow Ostankino TV First Channel, Feb. 5, 1995); and the strike at the Glushko plant Energomash (*Trud*, Dec. 10, 1994, p. 2).

Information from visits by NASA experts to Moscow came to me almost continuously. The quotes used here are primarily from a lunch meeting with an old friend on Nov. 8, 1994, which I wrote a description of within hours of the meeting.

The Albats story about meeting an Iranian-bound rocket scientist at a Moscow airport was from "If You Really Want to Worry, Think Loose Nukes," *Newsweek*, May 25, 1998, pp. 32, 32A. Congress asked Deputy Secretary of State Strobe Talbott about such missile technology export, and he sent a written response dated Oct. 7, 1998. U.S. ambassador to Russia James F. Collins' admissions were in response to a Feb. 2, 2000, CIA report to Congress. The subject was also described in David Hoffman, "Idled Arms Experts in Russia Pose Threat," *Washington Post*, Dec. 28, 1998, p. A01.

Chapter 16: *ISS* Opens

The game of "space chicken" in the fall of 1998 is adapted from my article "The U.S.-Russian Space Relationship: Symbolism at Any Cost?" in *Spectrum* magazine, July 1999, pp. 74–81. The discussion of the looming last-minute technological challenge of launching the 3A/4A/5A assembly missions on schedule is adapted from my article "NASA's Big Push to the Space Station," in *Spectrum*, November 2000. Both articles are based on contemporary press accounts, NASA press conferences and interviews (particularly with Tommy Holloway and Robert Cabana, both of whom were candid and cooperative), NASA *ISS* documents such as status reports and training manuals that happened to fall into my hands by various routes, and two dozen inside sources, all of whom need anonymity to avoid NASA retribution.

Chapter 17: Last Days of *Mir*

Quotations from pro-*Mir* cosmonauts are from contemporary press accounts. Grigoryev's comments were made on NTV television, Apr. 12, 1999. Kaleri's quote is from "Mourning *Mir*'s Demise,'" *Moscow Times*, Mar. 16, 2001. Manarov's and Avdeyev's quotes are from the *Los Angeles Times*, in an article by John Daniszewski, Moscow correspondent, Mar. 22, 2001.

Truly's report to Congress with the pessimistic (and in hindsight very accurate) assessment of *Mir*'s science potential was dated Oct. 23, 1991, and included the attachment entitled, "NASA Assessment of Current Soviet *Mir* Space Station and *Soyuz* Spacecraft."

Merbold's briefing to NASA was recorded in an internal memo by NASA engineer John Graf on Dec. 20, 1994, and I obtained a copy (not from Graf). Portree's assessment was conveyed to me in a long e-mail essay in February 2000. Cornett's report was in a private communication on June 23, 1999. The doctor-astronaut who said that "scientific goals" were adjusted to meet actual accomplishments also conveyed this to me privately. The 1994 NASA scientist comment on "concern" is from the Mar. 21, 1994, issue of *Space News*.

Various 1997 NASA memos described how scientific activities were being downgraded, or even abandoned. Nygren's June 24, 1997, e-mail and astronaut Wendy Lawrence's bitter reply are in my files. NASA's role in pressuring Russia to dump *Mir* was widely reported in contemporary press accounts. Brinkley's doubt about Russia's ability to handle both *Mir* and the *ISS* was expressed in a face-to-face interview at the Nov. 13, 1998, press conference.

Technical details on how Russian engineers would control *Mir* after the crew left in September 1999 were provided in personal communications from Dutch space radio listener Chris van den Berg and Russian space journalist Igor Lissov. Details on how the Russians attempted to keep funding *Mir* are from contemporary press accounts, as is the story of MirCorp's temporary rescue and NASA's vehement reaction to it. In particular, Goldin's comments were from congressional testimony on Apr. 6, 2000. Quotations from Representative Dana Rohrabacher are from wire stories the following day.

Details of how Russia kept *Mir* open for another year are adapted from my article "Saving *Mir* with a Rope Trick," in *Spectrum*, July 2000, which relied upon both contemporary press accounts and interviews with MirCorp, NASA officials, and several private individuals involved in the dealings, who requested anonymity.

The final months of *Mir* are chronicled from contemporary press accounts and private discussions with orbital experts within NASA. See also Rex Hall, ed., *Mir: The Final Year*, British Interplanetary Society, London, 2001. I am grateful to Miles O'Brien at CNN for organizing the "Mir Death Watch" special program and for giving me such a key role in it.

Chapter 18: Day-by-Day in Orbit

These accounts are adapted from the Ship's Logs written almost daily by commander Bill Shepherd, from in-flight press interviews which I watched, and from the transcripts of the postflight debriefings. I am grateful to space workers who provided me access to the original log entries and to the debriefing materials, all of which NASA is keeping secret.

Chapter 19: Future Orbits

The account of the artificial NASA/Russia flap over "space tourist" Dennis Tito is from contemporary press accounts and TV interviews, including the Larry King show. Readdy's post-TV comments were via private e-mail with me.

The hypocrisy about "safety concerns" was highlighted by the contents of the minutes of a Safety Review Panel at the Johnson Space Center on Feb. 28, 2001. I received a copy of those in-house minutes and quoted from that document. The Aug. 17–20, 1999, U.S.-Russia meeting on acoustic levels was documented in a contemporaneous "Weekly Activity Report" by one of the participants, and somebody on his mailing list sent me a copy. Actual acoustic levels on the *ISS* were discussed at NASA's private postflight crew debriefings in March 2001, and reports of those meetings were shared with me by participants. The same happened with written progress reports from the second expedition, which were unofficially shared with me even as NASA press officials denied that they existed.

Goldin's adulation of Russian space superiority over his own NASA teams can be seen on the private *NASA Watch* page (http://spaceflight.nasa.gov/history/shuttle-mir/mir23/status/week19/goldin.html). For balance, a good overview of the ingenuity of NASA's *Skylab* teams can be found in Henry Cooper's book *House in Space*, Holt, Rinehart and Winston, NY, 1976.

NASA's official *ISS* vision statement about the partnership as a goal in itself was posted on NASA *ISS* home pages in March 2001. Bryan Burrough's paean to a *Star Trek* universe was in the *New York Times*, Nov. 1, 2000. Bob Cabana, the astronaut who commanded the first space station assembly mission in December 1998, voiced an almost theological passion for a permanent space partnership in a radio discussion with me and John Pike on the Diana Rehm talk show (NPR) on June 7, 2000.

At a pre-crew-launch press conference at Baykonur, Michael Foale asserted that "the model for space exploration is international cooperation." His strategy: "This flight is the keystone to all future exploration from this planet—to the Moon, to Mars and asteroids." Former astronaut Mike Baker, now a NASA official, agreed: "From now on, I think that all of our endeavors in space, human endeavors, will be joint." (Both were quoted by the Associated Press, Oct. 31, 2000.) Malenchenko's quote is from *Florida Today*, Sept. 20, 1998. Fabian's quote is from his testimony before the House Science Committee, October 1993. Nicholas Burns was quoted by Interfax, Aug. 7, 1998.

Goldin's fatuous quote about "instead of pointing missiles at each other" was in *Aerospace Daily*, Nov. 1, 2000. The "better than building bombs" quote from Bowersox was in Reuters, Oct. 31, 2000, and in Scott Peterson, "Space Station Launches Friendships," *Christian Science Monitor*, Nov. 1, 2000. The Dieter Andresen quote was in *Florida Today*, Sept. 20, 1998. The well-argued essay by Robert Davies on future joint projects was "Space Station Is Worth It," *Christian Science Monitor*, Mar. 15, 2001.

The Biopreparat affair is described in the *Washington Times*, Oct. 6, 2000, and in *Space News*, Oct. 30, 2000. The "outrageous" comment from a senior U.S. national security official was in an article by Judith Miller in the *New York Times* on Jan. 25, 2000. I also received valuable insights into NASA's own "foreign policy" in the form of deliberately bypassing U.S. government export controls from a source at NASA Headquarters in an e-mail on Jan. 28, 2000.

The *New York Times* editorial repeating the old nonsense that the *ISS* would protect us from missile-armed "rogue states" was published on Nov. 1, 2000.

Koptev's comment about "gaining access" was in an interview in *Rossiyskaya Gazeta* on Nov. 26, 1998. His financial descriptions were at a press conference on Nov. 25, 1998, for which I have the transcript.

Kranz's more optimistic view was expressed in a letter in *Space News*, Apr. 22, 2001. Anatol Lieven's quote is from a column by Georgie Anne Geyer in the *Chicago Tribune*, Mar. 30, 2001.

Name Index

Subject Index

About the Author

James Oberg was a space engineer for 22 years in NASA's mission control in Houston. He has been the space consultant for ABC News, United Press International, and several foreign networks. Oberg is widely recognized as a world authority on the Russian space program, and he has been invited on several occasions to testify before Congress on the problems facing the Russian space industry. He is the author of 10 books, including the classic, *Red Star in Orbit*, and about a thousand magazine and newspaper articles on all aspects of space flight.